ADVANCE PRAISE FOR

Ecocriticism

"Gracefully written and solidly researched, *Ecocriticism* surveys the emerging field of ecocriticism and places Donelle N. Dreese among the growing number of critics who question dualistic discourses that separate humans from the natural world. Writing with keen insights honed from ecofeminist and postcolonial theories, Dreese focuses on American Indian literatures to illuminate the struggle writers face as they attempt to integrate the past with the present and reconcile their mythic/historical sense of place with their contemporary sense of place. The value of Dreese's fine book lies in its call for each of us to seriously reconsider those cultural values that are destructive to the earth. In short, this is required reading for all those interested in reimagining human relation to other species and to our communal and natural environments."

—*Joni Adamson, Associate Professor and Head of the Department of English and Folklore, University of Arizona; Author of* American Indian Literature, Environmental Justice, and Ecocriticism: The Middle Place *(2001)*

"In *Ecocriticism*, Donelle N. Dreese cogently explains the value of employing the concept of 'reterritorialization' in reading a variety of contemporary writers: Anzaldua, Berry, Chrystos, Griffin, Harjo, Hogan, Momaday, Ortiz, Rose, and Vizenor. She shows that white and native, male and female, poet and prose writer, share a concern for recovery, psychically and physically, through relocating and placing themselves and their characters. Dreese presents her theoretical concepts clearly and shows their efficacy through invigorating and stimulating readings across literary genres that demonstrate how all these writers are united by a sense of urgency about the need to reconfigure and defend their home territories."

—*Patrick D. Murphy, Chair, English Department, University of Central Florida, Orlando; Founding Editor,* ISLE: Interdisciplinary Studies in Literature and Environment

Ecocriticism

American Indian Studies

Elizabeth Hoffman Nelson and Malcolm A. Nelson
General Editors

Vol. 15

PETER LANG
New York • Washington, D.C./Baltimore • Bern
Frankfurt am Main • Berlin • Brussels • Vienna • Oxford

Donelle N. Dreese

Ecocriticism

Creating Self and Place in Environmental and American Indian Literatures

PETER LANG
New York • Washington, D.C./Baltimore • Bern
Frankfurt am Main • Berlin • Brussels • Vienna • Oxford

Library of Congress Cataloging-in-Publication Data

Dreese, Donelle N. (Donelle Nicole).
Ecocriticism: creating self and place in environmental and
American Indian literatures / Donelle N. Dreese.
p. cm. — (American Indian studies; v. 15)
Includes bibliographical references and index.
1. American literature—Indian authors—History and criticism.
2. Environmental literature—United States—History and criticism. 3. American
literature—History and criticism. 4. Environmental protection in literature.
5. Environmental policy in literature. 6. Wilderness areas in literature.
7. Landscape in literature. 8. Ecology in literature. 9. Indians in literature.
10. Nature in literature. 11. Self in literature. I. Title. II. Series.
PS153.I52 D74 810.9'897—dc21 2001029724
ISBN 0-8204-5661-6
ISSN 1058-563X

Die Deutsche Bibliothek-CIP-Einheitsaufnahme

Dreese, Donelle N.:
Ecocriticism: creating self and place in environmental and
American Indian literatures / Donelle N. Dreese.
–New York; Washington, D.C./Baltimore; Bern;
Frankfurt am Main; Berlin; Brussels; Vienna; Oxford: Lang.
(American Indian studies; Vol. 15)
ISBN 0-8204-5661-6

Cover design by Dutton & Sherman Design

Contents

Contents

Acknowledgments

I have been very lucky while writing this book to have been surrounded by good friends and colleagues. My debts are deep, but none so great as to Patrick Murphy, who introduced me to this marvelous literature, which inspired me and changed my life. As a scholar, a teacher, and as a person, he has my highest respect and regard.

I would like to thank Susan Comfort and Judith Villa, who read through the earliest manuscripts, and whose guidance led me toward further inquiry about my topics and positioned my work into a larger field of scholarship.

I would also like to thank Jackie Pavlovic and Phyllis Korpor for their gracious enthusiasm for the project, and for their support and patient guidance through the production process.

I'm pleased to offer generous gratitude to the scores of talented students and colleagues who, through their genuine curiosity and impassioned conversation, never failed to pour inspiration into the energy that brought this project to completion.

There are many other special people to whom I am indebted. It is a privilege to name them here and to express my appreciation for their love, friendship, advice, support, and encouragement: Michael F. Gaynord, John-Patrick Driscoll, Malcolm Hayward, Cecilia Rodríguez Milanés, Fred Jordan, Stephanie Dowdle, Dong-oh Choi, Stephen Housenick, Nathan Morgan, Nick Mauriello, Elizabeth Byrne, Lisa Blair, Judith Newlin, Chris Harlos, Virginia Silva, Meredith Sykes, Chris Cobb, Lucindy Willis, Joni Adamson, and Ben Williams.

Grateful acknowledgment is hereby made to copyright holders for permission to use the following material:

From *Fire Power*, by Chrystos. Press Gang Publishers, Copyright © 1995. Reprinted by permission of the author.

From *Not Vanishing*, by Chrystos. Press Gang Publishers, Copyright © 1988. Reprinted by permission of the author.

From *The Woman Who Fell from the Sky: Poems.* Copyright © 1994 by Joy Harjo. Reprinted by permission of W. W. Norton & Company Inc. and Joy Harjo.

Hogan, Linda. *The Book of Medicines.* Minneapolis: Coffee House, 1993. Reprinted by permission of the publisher. All rights reserved.

Momaday, N. Scott. *The Way to Rainy Mountain.* Albuquerque: University of New Mexico Press, 1969. Reprinted by permission of the publisher. All rights reserved.

Rose, Wendy. *Bone Dance: New and Selected Poems, 1965–1993.* Tucson: University of Arizona Press, 1994. Reprinted by permission of the publisher. All rights reserved.

Chapter One

Ecocriticism, Sense of Place, Reterritorialization, Postcolonialism, and Nostalgia

> Another question is raised: is not the purpose of all this living and studying the achievement of self-knowledge, self-realization? How does knowledge of place help us know the Self? The answer, simply put, is that we are all composite beings, not only physically but intellectually, whose sole individual identifying feature is a particular form or structure changing constantly in time. . . . Thus, knowing who we are and knowing where we are are intimately linked. There are no limits to the possibilities of the study of *who* and *where.*
>
> —Gary Snyder, *A Place in Space*

Wendell Berry once made the statement that in order to know who you are you must first know where you are. Whether we are cognizant of their influences or not, environmental factors play a crucial role in the physical, emotional, and even spiritual configurations that determine our ideas of who we are. All human beings develop their own sense of place through life that determines why they love certain regions or feel utterly alien in others. It is not an uncommon human experience to long for the particularities of a certain place that have had a powerful interior effect on their human psyche. Neil Evernden, for instance, observes that

> there appears to be a human phenomenon, similar in some ways to the experience of territoriality, that is described as aesthetic and which is, in effect, a "sense of place," a sense of knowing and of being part of a particular place. There's nothing very mysterious about this—it's just what it feels like to be home, to experience a sense of light or smell that is inexplicably "right." (100)

While there may be nothing mysterious about this phenomenon, I'm not sure how many of us seriously consider it and recognize how powerful this pull can be toward what feels "right" or like "home." The following chapters will attempt to demonstrate the complexity involved in defining a sense of home and how it is connected to many other facets of being human.

After several evolving discussions of place occurred in four freshman/sophomore-level English writing courses, I reflected on student responses and found that the impact of place, at least among these eighty-four students, was more individual and unrecognized than I had imagined. Many students were completely unaware of the fact that they had been in any way affected by their environments, while few were acutely aware of its influence. Part of this lack of awareness of place may be due to the fact that many of the students were born in the area of the college/university and have done very little traveling outside the area for comparison. Perhaps some of the students had not yet had the opportunity to experience being in a place that feels uncomfortable or unlike home. The need for a basis of comparison when attempting to define home perhaps suggests that "a sense of place" requires boundaries and an identifiable notion of what is outside or beyond one's sense of place or home. It would support the tentative assertion that understanding the self requires an understanding of what the self is not; that exploration in both territories is perhaps necessary for a deeper self-comprehension. Regardless, place has made an impact, and one responds in accordance with that impact, even if it is unconscious.

Perhaps there is no place more influential in the development of the human identity than the place where one grows up. Individuals who have spent childhood years in an urban environment may feel most fully connected to themselves when they are surrounded by street noise, concrete, and the smell of gasoline exhaust, while those whose childhoods were immersed in a more rural setting may desire a natural environment where they feel most comfortable. If those experiences were particularly undesirable, negative responses may be evoked in the presence of childhood environments.

The sense of place within each of us is very sensual. It engages all of our senses on a daily basis until we may hardly be aware of what we see, smell, hear, or feel in the place we call home. But it is also highly mental and emotional. Perhaps local culture is something we all take for granted and neglect until we're out of its sphere.

I do contend, however, that we as human beings are engaged in the eternal search for connection, for that which connects us to others and for that which connects us to ourselves. Culture, language, history, belief systems, social practice, and other influences on human development are as much a part of place as the physical landscape one crosses. If that place which provides the connection we desire is not readily available to us, we find a way to create our own space or home, which we can inhabit and feel at ease with ourselves and our surroundings. Writers are certainly not exempt from this human search for connection, and the way that it manifests itself through their writing is at the core of this study.

Within the last few years, what was once considered literary regionalism as exemplified by such writers as Sarah Orne Jewett, Mark Twain, and Mary E. Wilkins Freeman has evolved into this study of the "sense of place." That title, now heading chapters in many texts for lower-level English courses, was proposed by Michael Kowalewski in his essay "Writing in Place: The New American Regionalism" in an effort to redefine the concept of regionalism by moving it away from its rural limitations to include such diverse landscapes as the urban and western. Kowalewski claims that "the time seems ripe to replace a definition of regions in which large urban-suburban portions of the map have been artificially removed with one in which the full spectrum of places within a given area . . . can be studied and described" (180). With this broader notion of place becoming more prominent, writers and critics are discovering new literary territories to explore. This project investigates place as a physical, psychological, ideological, historical, and environmental construct where writers challenge and alter these constructions in order to create a habitable place or home.

Working from postcolonial and ecocritical theoretical notions that place is inherent in configurations of the self and in the establishment of community and holistic well-being, the purpose of this book is to examine the centrality of landscape in contemporary poetry and prose works by writers who, either through mythic, psychic, or geographic channels, have identified a landscape or environment as intrinsic to their own conceptualizations of self. Questions that are asked of the texts chosen include: How does the author present the landscape, and what is his/her attitude toward it? What is the sociopolitical or ethical agenda, if any, of the author in writing about a certain place? How do the characters interact with their environment, and are there any conflicts present in that interaction?

Wendell Berry in *Home Economics* states: "When we propose that humans should learn to behave properly with respect to nature so as to place their domestic economy harmoniously upon and within the sustaining and surrounding wilderness, then we make possible a sort of landscape criticism" (151). This kind of respect for nature and an awareness of interconnectedness are probably the most basic tenets of the rapidly growing literary theory known as ecocriticism, or as Berry puts it, landscape criticism. Ecocriticism, a term first coined by William Ruekert in 1978, addresses issues concerning landscape and the environment that have previously been overlooked by the literary academy. A few examples would include: how nature is represented, when it is represented, how the environmental crisis has influenced literature, and how concepts of the environment have evolved through the centuries. Cheryll Glotfelty in her introduction to *The Ecocriticism Reader* states that "nature per se is not the only focus of ecocritical studies of representation. Other topics include the frontier, animals, cities, specific geographical regions, rivers, mountains, deserts, Indians, technology, garbage, and the body" (xxiii). Ecocriticism, therefore, covers a broad range of issues indeed, involving all that which comprises our human interior and exterior contexts. An important conviction of ecocriticism is that we are interconnected with the world around us and, therefore, studying the environment involves studying how human beings affect and interact with the environment.

Glen A. Love in his article, "Revaluing Nature: Toward an Ecological Criticism," questions the academy and its commitment to interdisciplinary studies and contemporary issues while still ignoring the environmental crisis and its threat to human survival. Ecocriticism, as an activist philosophy, has as one of its primary agendas the reduction of dualistic thinking that has separated the human being from the natural world in Western discourse and practice. Dualisms can only be reduced by first creating an awareness within the academy that this type of bipolar thinking only perpetuates destructive binary notions that have previously placed environmental concerns on the negative side of the dualism. In 1990, Love reproached his colleagues, saying:

> [T]he decision of those who profess English has been, by and large, that the relationship between literature and these issues of the degradation of the earth is something that we won't talk about. Where the subject unavoidably arises, it is commonly assigned to some category such as "nature writing," or "regionalism," or "interdisciplinary studies," obscure pigeonholes whose very titles have seemed to announce their insignificance. (203)

Fortunately, in the intervening years recent critical publications suggest that ecocriticism is gaining more recognition and that more academic positions devoted to environmental literature are being created. Ecocriticism has several related disciplines, such as deep ecology, ecofeminism, social ecology, and environmental justice. A brief discussion of several of these in this introduction would be useful due to their prominence in this literary exploration.

Annie Booth and Harvey Jacobs state that "Deep ecology attempts to examine the deeper root question concerning human interactions with the natural world, rather than the 'shallow' issues such as pollution or species extermination, which it identifies as more the symptoms than the cause of environmental breakdown" (29). Deep ecology rejects Enlightenment notions, which separate humans from nature, perpetuating an objectification of the natural environment. This objectification of the natural world serves as the justification for the continued exploitation and degradation that lies at the core of the global environmental crisis. Deep ecology challenges the hierarchy that has polarized humans and nature and advocates a biocentric perspective, which acknowledges the mutually reciprocal relationship required for a sustainable ecosystem.

SueEllen Campbell in "The Land and Language of Desire: Where Deep Ecology and Post-Structuralism Meet" compares post-structuralist literary theory with deep ecology to discover what is for her the most important shared premise between the two: "both criticize the traditional sense of a separate, independent, authoritative *center* of value or meaning; both substitute the idea of *networks*" (206–207). Under this mode of thought, humans, plants, and animals coexist on an equal sphere within a intimate system of connections where it is impossible for one part of the system to change without influencing and affecting another. One article states that "Deep ecology argues that all life on Earth from humans to ecosystems to soil microbes possess equal intrinsic value, values which exist independent of human needs and desires" (Booth and Jacobs 29). What complicates this essentialist view is that human needs and desires do have an impact on the environment that is astonishing, despite debates arguing the actual existence of an environmental crisis. Laws to control pollution are not up to pace with the ever-increasing population rate and rapidly expanding industrialization. Many activists contend that the future resides in each individual and that the choices we make reflect our values. Currently, there is a greater value for consumerism and development than a recognition of being an equal member within a delicate ecosystem, and those values are reflected in the decline of our wilderness lands and our high cancer rates. Deep ecology sees the scope of all environmental exploitation as symptomatic of a much deeper nature/human relational breakdown.

American Indian environmental philosophies have made a vital impact on the development of ecocriticism. The influence of these philosophies rests in their unparalleled ability to demonstrate conceptualizations of nature which, by their very contrast, hold a mirror to Western capitalist notions of commodification and require a re-evaluation of their practices in the presence of the recognized crisis. Booth and Jacobs affirm that many "American Indian cultures adapted their needs to the capacities of natural communities; the new inhabitants, freshly out of Europe, adapted natural communities to meet their needs" (31). This new inhabitant's pattern of thought concerning the environment established the relationship and attitudes many human beings would have of the landscape from the time of colonization to the present day.

Similarly, Gaia theory, which recognizes the earth as a living, conscious organism, introduced an ethical component into colonial and contemporary uses of the environment, calling into question the objectification necessary for abuse of the natural world. When British atmospheric scientist James Lovelock proposed his Gaia hypothesis in the early 1980s, it came as quite a shock to the scientific world that the earth could be viewed on such a global scale as a single living entity whose constituents function in order to maintain balance or homeostasis. The theory was apparently favorably received by indigenous commentaries because the hypothesis was not new to them. In *The Sacred Hoop: Recovering the Feminine in American Indian Traditions*, contemporary Laguna writer Paula Gunn Allen clarifies the notion of the earth as a living being among American Indian communities:

> An Indian, at the deepest level of being, assumes that the earth is alive in the same sense that human beings are alive. This aliveness is seen in nonphysical terms, in terms that are perhaps familiar to the mystic or the psychic, and this view gives rise to a metaphysical sense of reality that is an ineradicable part of Indian awareness. (70)

At the time of the Gaia theory's introduction, there were arguments among its followers as to whether it was a scientific theory or a theological one, indicating that hard scientific and spiritual approaches to studying the earth were at odds. From an American Indian perspective, I don't think this is an area of controversy because its life philosophies did not dictate such dichotomies. By acknowledging the earth as a living organism, alive in the same way human beings are alive, Gaia theory perhaps introduced an ethical consciousness into environmental studies of the Western world that have been in practice among indigenous cultures for centuries.

While viewing aspects of the environment with spiritual reverence as nothing less than kin and often as the embodiments of Gods or figures of great wisdom, most American Indian cultures have evolved from a tradition that cares for the landscape with respect and reciprocity. That which is taken is returned through prayer, ritual, and ceremony to maintain the delicate balance upon which all life rests. Abuse or poisoning of the land would inevitably lead to a disruption of that balance, which would, in turn, cause pain and suffering in physical and spiritual forms for the life which inhabits it. As Paula Gunn Allen has observed in *Grandmothers of the Light: A Medicine Woman's Sourcebook*, "it is the loss of harmony, an inner-world imbalance, that reveals itself in physical or psychological ailment. It also plays itself out in social ailments, war, dictatorship, elitism, classism, sexism, and homophobia" (168). Realistically, it has been historically inconceivable to Western modes of thought to suggest such an all-encompassing connection between the environment and the state of human existence as viable. Ecocriticism, however, much like some American Indian philosophies, promotes and teaches the interdependence and connectedness of all living things, which means that any study of human existence would be insufficient if it did not place us within an environmental context.

The sense of place and its relationship to the self are perhaps nowhere more evident than in American Indian literatures. Writers such as Louise Erdrich, Joy Harjo, Paula Gunn Allen, James Welch, N. Scott Momaday, Linda Hogan, and Leslie Marmon Silko are just a few of the contemporary American Indian writers who explore the theme of Native identity and place in their works. The roots of the identity of a tribe are initially determined by the oral tradition creation and emergence stories. These stories tell of the ancient peoples of a tribe, how they came into being and how they emerged, usually from some form of darker underworld through a hole, or a spring, or a log, into the world of light and earthly habitation. These stories and others within the oral tradition describe specific landscapes from which a tribe derives its means for survival, its cultural symbols, its sense of self, and its spirituality.

In "Landscape, History, and the Pueblo Imagination," Leslie Marmon Silko describes this centrality of landscape and place within the oral tradition:

> As offspring of the Mother Earth, the ancient Pueblo people could not conceive of themselves without a specific landscape. Location, or "place," nearly always plays a central role in the Pueblo oral narratives. Indeed, stories are most frequently recalled as people are passing by a specific geographical feature or the exact place where a story takes place. (269)

The oral tradition, which sustains a tribe's cultural identity through historical record, documents not only the creation and emergence of a tribe but also migration, animals, or sites with particular symbolic or spiritual significance. This oral historical record chronicles how daily lives were affected by the conditions of the landscape, and how a tribe relies on elements of the earth for shelter and sustenance. Clearly the diversity of the landscape of what we now call North America has contributed greatly in creating a diversity of American Indian cultures, sharing some fundamental ideologies of worldview but ultimately demonstrating a highly pluralistic life existence in practice and worship.

It is important to discuss aspects of American Indian cultures in this introduction because my topics—landscape, sense of place, and identity—are integral in all human development but particularly in American Indian cultures and literatures. I hope it is clear that my purpose is not to speak for or deny agency to any of the writers I explore. My purpose is to further the study of the literatures of these writers, who through their artistry and wisdom deserve more attention than they have so far received.

Another discipline related to ecocriticism and also important to this study is ecofeminism. Upon first encountering this term, it would not be unusual for someone to ask, what does nature have to do with women, specifically? Why is there considered a special relationship between the two that would require its own theory?

In simplest terms, ecofeminism draws comparisons between the oppression and domination of nature with the subjugation of women. Nature and women, considered dangerous and unpredictable, need to be subdued and controlled. If we understand the domination of nature as that which exploits, devalues, destroys, and/or renders powerless natural resources, ecofeminism states that the same kind of philosophy behind this domination is inherent in that which has had the same effect on women. What this means, according to Greta Gaard, is that "the standard history of colonialism, one in which the oppressive structures of capitalism, Christianity, and patriarchy construct nature and in which those associated with nature are considered resources for the colonizer . . . interesting only in terms of their subordination" (12). The alienating and destructive dichotomies nurtured by Western metaphysical ideologies are what ecofeminists are trying to dismantle. Culture/nature, mind/body, black/white, man/woman, intellect/emotion are all examples of structures that lie at the root of subordination and are perpetuated by those who benefit from them. Because the ecofeminist agenda involves healing these artificial separations and challenging existing power structures, writers and theorists such as Susan Griffin and Ynestra King, among many others, have made writing an activist endeavor to help us better understand ourselves and make connections where there are gaps.

Perhaps the newest term or movement related to ecocriticism is environmental racism. The term was first coined in 1982 in Warren County, North Carolina, during a rally protesting a PCB landfill that was scheduled for the area, which was predominantly rural, poor, and African American. Benjamin Chavis's definition of environmental racism is direct and explicit:

> Racial discrimination is the deliberate targeting of communities of color for toxic waste disposal and the siting of polluting industries. It is racial discrimination in the official sanctioning of the life-threatening presence of poisons and pollutants in communities of color. And it is racial discrimination in the history of excluding people of color from the mainstream environmental groups, decision making boards, commissions, and regulatory bodies. (3)

Although the term may be relatively new, the practice certainly is not, as Kamala Platt reminds us in her article "Ecocritical Chicana Literature" in the summer 1996 issue of *ISLE*. The beginnings of environmental racism can be dated to the onset of European colonization and later when American Indian tribes who once inhabited life-sustaining environments were removed to landscapes far less fertile or lacking in the natural resources necessary for survival (69).

While it has perhaps taken too long for terms such as environmental racism, deep ecology, and ecofeminism to emerge into the field of English studies, the growing enthusiasm, job market, and number of publications with ecocritical themes do suggest that English studies is stepping outside and becoming part of the environmental justice movement. Ecocritics argue that the environmental crisis should be a concern as prominent in English classes as studies of class, race, and gender issues. I would tend to agree. Its delay in getting there simply further exemplifies the human alienation from nature that lies at the core of the environmental crisis.

The various environmental movements surveyed in this introduction are critical orientations that provide environmentally conscious avenues of exploration for the works of literature that I study. I have discussed them here to support and present background for my theoretical perspectives and interpretive approaches, and to clarify for readers the importance of my study. Although deep ecology is not directly discussed in the following chapters, the forthcoming analyses give credence to its tenets—that environmental problems are symptomatic of an underlying crisis. And it is only by identifying the root causes that we will be able to propose a solution. These reviews of ecofeminism and environmental racism are meant to link women and "minority" cultures with the environment and landscape as members of a community that share a history of oppression. The identifiable history of exploitation and misuse experienced by women and American Indians in relation to the land is the basis for many of my discussions and allows the different environmental movements to complement one another for a common cause—to improve the way human beings behave towards the places they inhabit and one another. Summations of theoretical concepts and literary movements will be highlighted in the conclusion.

~~~~~

If self-knowledge and self-realization are simultaneously threaded in the fabric of where, as Gary Snyder suggests in the quote that opens this chapter, then a sense of place would be a key component in the study of identity politics and representation of the self in literature. In "Deterritorializations: The Rewriting of Home and Exile in Western Feminist Discourse," Caren Kaplan explains deterritorialization as a "term for the displacement of identities, persons, and meanings that is endemic to the postmodern world system" (358). I would add to this the postcolonial world system as well, where physical, cultural, and spiritual displacements of non-Western cultures by colonial forces have left and still leave these cultures uprooted and removed from their cultural heritage. Postcolonial thought, essentially, has emerged as a response to continuing racial oppressions and contemporary global conditions involving issues such as the effects of the colonial legacy, cultural hybridity, and plurality. Interests in this theory stem from the movement of literary studies into a more diverse, multicultural discipline reflecting a more multiculturally conscious academe.

While postcolonialism involves many positive ideals, it has been difficult to define and reconcile with its origins. Anne McClintock, in her essay "The Angel of Progress: Pitfalls of the Term Post-Colonialism," addresses the problem with the term
~~~~~

"postcolonialism" and why there is a need for terminology that is more specific and historically accurate. The prefix "post," as she indicates, provides a misleading suggestion that colonialism existed in the past and that we are now in an era which has shed this system of power. She reminds us of countries that are still under colonial domination today while pointing out that the prefix is one that follows along a linear construct of time that is part of Western philosophy and therefore works against the kind of ideologies that postcolonialism is trying to dismantle. "Metaphorically, the term 'post-colonialism' marks history as a series of stages along an epochal road from the 'pre-colonial,' to the 'colonial'—an unbidden, if disavowed, commitment to linear time and the idea of 'development'" (292). She believes that the "post" is a symptom of a larger global crisis in which ideological and ecological foundations are breaking down and presenting the preview of an apocalyptic future. Although this vision of the future is bleak, it provides a provocative explanation for the difficulty the present age has in defining itself.

Vijay Mishra and Bob Hodge also address the meaning of the term postcolonialism and its usefulness in addressing the First and Third World relationship in literature. Although Mishra and Hodge agree with McClintock that the term suggests a periodicity which has already passed, they do acknowledge that it provides a door by which to enter the discussion and dialogue of colonial power and its effect on Third World countries. The theory and term foreground "a politics of oppression and struggle, and problematize the key relationships between center and periphery. It has helped to destabilize the barriers around "English literature" that protected the primacy of the canon and self-evidence of its standards" (276). They go on to define two kinds of postcolonialism that they find most appropriate in defining its dynamics in literature. One, "oppositional postcolonialism," presents a direct, overt form of resistance to colonial constructs, and the other is what they term "complicit postcolonialism."

Complicit postcolonialism is a problem for Mishra and Hodge because of its apparent relationship with postmodernism. In their article, postmodernism is presented negatively. They feel that if a postcolonial text bears characteristics of the postmodern novel that it will subvert its message, which is not the agenda of a postcolonial text. I disagree with this view, however, and do examine a fiction work, which is considered postmodern in this investigation, Gerald Vizenor's *Bearheart: The Heirship Chronicles.* Vizenor's text, in all its postmodernity, draws attention to the postcolonial agenda rather than subverting it by using bizarre characters and situations that only illuminate the cultural and ecological disasters that are materializing rather than hiding them.

In addressing American Indian concerns, one begins to wonder where the United States fits into the postcolonial concept. Amy Kaplan, in an article entitled "'Left Alone with America': The Absence of Empire in the Study of American Culture," makes a few important points about academic postcolonial theory. She states that there has been a kind of American exceptionalism in the study of imperial power in the United States involving three things: denial, displacement, and projection. The United States denies that it is part of the imperial process by claiming that it has historically been a victim of British imperial power. By displacement Kaplan means that instead of using imperialist language to describe itself, American academic institutions use terms such as "discovery" and "exploration," which have positive connotations and evoke the spirit of adventure as opposed to images of territorial appropriation. By projection, she refers to looking at individuals such as Saddam Hussein as colonial perpetrators instead of giving an internal critique of their own imperialistic agendas.

This aspect of American exceptionalism is significant in that it describes the dominant culture's attempts to mask or eliminate itself from imperial endeavors. She explains that

> United States continental expansion is often treated as an entirely separate phenomenon from European colonialism of the nineteenth century, rather than as an interrelated form of imperial expansion. The divorce between these two histories mirrors the American historiographical tradition of viewing empire as a twentieth-century aberration rather than as part of an expansionist continuum. (17)

Kaplan's article draws attention to the colonial oppression that has taken place in the United States and discusses it as being as devastating to marginalized cultures as the colonialism in other parts of the world such as India and Africa. The author also perceives American Indian removals as part of the Industrial

Revolution and colonial project, which are recurring themes in American Indian literature and this analysis.

Amy Kaplan's essay also forces us to ask where postcolonial theory resides in the system of knowledge that created it. Arif Dirlick, in an article entitled "The Postcolonial Aura: Third World Criticism in the Age of Global Capitalism," states that postcolonial theory is a production of imperial power, meaning British and United States academics, and is therefore self-serving in its agenda. Dirlick asks the important question: what does postcolonial theory do for the Third World countries that it examines? I think that is a valid question. Does it actually help these countries and peoples under domination, or is it only there to provide intellectual discussion for American academics? Was it created in order to ultimately reinforce and validate their own position of power over Third World countries?

Chandra Talpade Mohanty believes that Western feminists define Third World women as bipolar opposites—uneducated, submissive, sexually restrained, traditional—of the self-defined Western feminists, who are strong, independent, educated, and in control of their lives. Mohanty criticizes the lack of cultural specificity in Western feminism and in postcolonialism in general and argues against looking at Third World women as subjects for analysis from an impersonal, imperial viewpoint.

Anne McClintock might agree with Dirlik as well in that she mentions postcolonialism's popularity in terms of its academic marketability. I would however, argue that it is a worthwhile theory that helps to counteract ideologies and practices that continue to benefit a white European center. Whether it is popular because of academic marketability is debatable; however, postcolonialism has surfaced at this point in time in history because there is real need for it. Scholars have become more aware of "minority" rights with the rise of the multicultural movement, and postcolonial theory fosters that awareness. The multicultural movement has given rise to the need for a deeper study of colonial forces and history. Also, many of the practitioners of postcolonial theory are displaced intellectuals from colonies and former colonies, such as Stuart Hall, Arjun Appadurai, Rey Chow, Homi Bhabha, and others. To avoid reasserting the center in my own work, I explore literature by writers of diverse ethnicities, support my ideas with theorists who represent diverse cultures, discuss the cultural histories of the writers, and dismantle Eurocentric ideologies and practices that have become sites of social, cultural, and environmental activism for these writers.

Finally, postcolonial theory, like everything else, is culturally situated and not without its biases. However, the discussion of the nature of postcolonial theory is perhaps more important than any conclusions we may draw. It is a significant field of study because it creates a much greater understanding of the concerns and desires of "minority" cultures that were once overlooked due to unconscious biases. With this postcolonial consciousness, educators and scholars can become more sympathetic and empathetic in their own cultural awareness, and how they read literature is a direct reflection of that greater sensibility. My approach and theoretical perspective hopefully mirror this sensibility by not privileging any one particular center in my readings of the works of literature.

Part of the postcolonial condition is a loss of the self, a cultural alienation involving an eradication of cultural traditions, a history, and national character. A response to the alienation is the attempt by colonized cultures to retrieve and reestablish a sense of cultural identity. Perhaps this would be a good time to discuss the notion of reterritorialization, whether or not it is actually viable, or if the retrieval of a lost identity is merely a nostalgic reflection and imaginative construct.

According to Stuart Hall, in "Cultural Identity and Diaspora," identity is ever-changing, a dynamic process unable to return to a pure prior state. After cultural syncretism has occurred, national character is forever altered; a door has been passed through by which one can never return. Hall believes that any attempt at re-establishing a historical identity is ultimately symbolic rather than true. He discusses how cultures attempt to revitalize their national identity through the production of film because of the industry's mass availability, and how the cinematic forum then becomes the basis for a new collective identity. The genre of film, however, bears the characteristics of imaginative production, and therefore indicates that the foundation upon which these cultures are envisioning themselves is one that is imaginatively based rather than any kind of actual resurgence of an historical essence.

Nationalism, Colonialism, and Literature (1990), by Terry Eagleton and others, agrees with Hall in its discussion of metaphysical essentialism. It claims that cultures that have undergone colonial transformation attempt to regain a lost essence but ultimately fail to do so. One of the problems he mentions is that metaphysical essentialism suggests that there is a true or natural national identity. Societies, therefore, run the risk of legitimating their culture over others, which is a reflection of the Eurocentric notion partially responsible for the success of the imperial process.

Arjun Appadurai, in "Disjuncture and Difference in the Global Cultural Economy," provides an interesting discussion about cultural flow and interchange that explains recent attempts by cultures to establish a coherent national identity. He asserts that imagination has become a social practice not only for Third World countries but for the United States as well. By listing the five dimensions of cultural flow, he suggests that these interchanges diminish a sense of structure and regularity among countries and result in people feeling nostalgic for pasts that they have never lived.

The five dimensions are media, technology, finance, ethnicity, and ideology—all of which he calls "scapes" to indicate the fluid nature of their dispersal. For example, he discusses a Philippine restaurant in the guise of a 1950s American malt shop scene, which produces a nostalgic and imaginative concept of identity that has never actually been experienced by the people of that country. He states that this cultural flow can cause breakdowns in the family if family members receive information from areas around the world that evoke changes in ideological practices at home or encourage significant relocation for employment. He uses the example of a person from a family built on strong religious affiliation who adopts a religion from another part of the world, thus alienating himself from his family structure.

The concept of re-establishing an identity through imagination is an interesting one and a recurring theme in postcolonial literature. Whether by nostalgic means or imagination as a social practice, retrieval of lost identity is a potent part of the postcolonial condition. What all the theorists presented here seem to conclude is that any prior state of cultural consciousness is gone once colonization has occurred, and one can only look back in nostalgia at what was lost or attempt to creatively reconstruct life anew. How then do we address the fundamental human desire to create connections, to have a history, to be a member of a cultural community? How can we resist becoming homogenized within a world with such intense global interchange? I do not provide concrete answers to these questions, but nostalgia and imaginative constructs are manifestations of their complexity.

Where does the concept of reterritorialization and the landscape fit into the colonial legacy? Many twentieth-century poets and prose writers of diverse ethnicity have attempted to recover a sense of home, identity, community, and place in response to various forms of displacement caused by colonization, oppression, and environmental alienation. One example is what I will refer to as "mythic reterritorialization," which occurs in writers from different American Indian cultures where oral traditions are prominent and the retrieval of a sense of origin and place is central to constructions of identity. As stated earlier, part of the postcolonial condition is a sense of emotional and physical dislocation involving a loss of the self, a cultural alienation involving an eradication of cultural traditions, history, and national character. Along with this dislocation comes the attempt by colonized cultures to retrieve a sense of cultural identity.

Much postcolonial criticism argues over whether the retrieval of a lost identity is actually viable or merely an imaginative construct, as Arjun Appadurai suggests, or symbolic, as Stuart Hall suggests. Joy Harjo, in *The Woman Who Fell From the Sky*, for example, retells this traditional Native tale by having the woman fall into an urban environment of tin cans and alley cats in order to be caught by her war veteran love interest. In this manner, Harjo imaginatively reconstructs a story from the oral tradition in order to appropriate its meaning to her own modern life and terrain. Harjo's poetry also inhabits a territory that lies in a liminal space between the mythic and mundane, which gives much of her later poetry a quality of mysticism residing on a metaphysical frontier.

Writers such as Linda Hogan in *The Book of Medicines* and N. Scott Momaday in *The Way to Rainy Mountain* evoke territories reminiscent of tribal origins, such as the Kiowa emergence from the log and the Chickasaw origin in water. In Hogan's *The Book of Medicines*, for example, water becomes a recurring image for physical and psychological healing in a contemporary world of drought and hunger. *Rainy Mountain*, with its sections broken up into a three-part structure of legend, history, and present reflection, reveals Momaday's purpose of relating himself to his Kiowa ancestors by showing the relationship between the tribe's mythic origins and their actual historical experience of living on the plains. Hogan, Harjo, and Momaday are all attempting to integrate the past with the present in their works to reconcile their mythic/historical sense of place with their contemporary sense of place.

Reterritorialization is not a solely physical act but one that is deeply emotional and intellectual as well. These "psychic reterritorializations" occur when landscape or place is used metaphorically to represent sites of conflict or refuge where writers more closely examine borders and zones of human and ideological contact. Using primarily the work *Not Vanishing*, I examine the poetry of Chrystos, using Bakhtinian dialogics, to describe how she presents her text as a battlefield, comprised of enemies and allies, where she fights battles with dominant culture oppression and finds refuge on a floral terrain with lesbian women of color. Theorists Vijay Mishra and Bob Hodge would agree that the poetry of Chrystos in *Not Vanishing* exemplifies their form of "oppositional postcolonialism." In this work, Chrystos's goal is to be as direct and clear as possible in expressing the actual material conditions of contemporary American Indian life. One poem entitled "No More Metaphors" expresses her view that when speaking about subjects such as racism, sexism, poverty, and violence, there is no room for metaphors. She believes that these issues have been buried long enough by a dominant power that would prefer to leave them unacknowledged. In the overtly political poems in her text, Chrystos makes a point of speaking so directly that her poems are painful to read. She adopts a voice of documentary realism and bares the harsh life of American Indians for the world to see.

Using poststructuralist and ecofeminist approaches, I explore Gloria Anzaldúa's *Borderlands/La Frontera*, as discussed earlier, and Susan Griffin's *Woman and Nature*, which both expose patriarchal and Western cultural constructs in order to remap their own terrains in order to break down dualistic thinking. For Anzaldúa, this involves the creation of the "New Mestiza," her term for a border consciousness that is safe for her inhabitation. Anzaldúa reconstructs her identity imaginatively. In this work, she writes about the difficulties of border life and how being of Mexican, Indian, American descent, along with being a woman and a lesbian, leave her open to rejection from all sides of her heritage. She is rejected from her Mexican heritage for being lesbian, and from the white heritage for being a woman of color. In order for her to locate a territory safe to inhabit she creates the new mestiza concept/image, which is a blend of all three heritages and the aspects therein that she chooses to accept. Her most pronounced argument in the book is that there is a desperate need for a massive uprooting of dualistic thinking, which sets people up against one another whether in terms of gender, religion, or culture. The new mestiza is a cultural configuration where there is acceptance for all types of individuals, a place where growth occurs and where identity is mutable and transformative.

Griffin, in the first half of her book, provides an elaborate ecofeminist study of how women have been deterritorialized through patriarchal oppression. By the end of the work, she reclaims the natural world and all of its inhabitants as a means of empowerment. By identification with nature rather than the alienation proposed by Western and patriarchal constructs, Griffin presents a utopian terrain for women to redefine themselves.

In a more literal sense, environmental reterritorialization involves writers who position themselves in natural settings in order to reinhabit a landscape or place that is intrinsic to their philosophies of being in the world. Reterritorialization, as described by Caren Kaplan, takes place when "we reinhabit a world of our making" (365), creating a terrain which is "a space in the imagination which allows for the inside, the outside, and the liminal elements in between" (367), according to the boundaries one sets. Gary Snyder believes that this reinhabitation refers to the defining of oneself in relation to place and recognizing the interconnectedness of life within the ecosystem. Writers who position themselves in specific terrains in their work, such as in nature writing or travelogues, foreground the importance of the relationship between the environment and the self and use the experience for self-examination and growth.

Linda Hogan in *Dwellings: A Spiritual History of the Living World* has a similar approach to nature in that she seeks to redefine Western culture stigmas that have been placed on certain creatures such as bats, snakes, and bees while also examining the inhabitory existence of these creatures to demonstrate that human beings are not separate from the natural world. She emphasizes the importance of place throughout this work of creative nonfiction and interweaves American Indian storytelling and philosophy in her writing.

Reterritorialization can also describe attempts to redefine notions of a particular landscape or refer to the desire to adjust one's natural environment in order to emphasize the relationship between the natural world and the self as reciprocal and historical, such as in Wendell Berry's *Clearing*. Some writers demonstrate relocation and travel as part of the search for self and community identification, such as Wendy Rose in her poetry and Simon Ortiz in *Woven Stone*.

Although all the writers that I have mentioned thus far are activists in their own right, several writers that I will discuss take a more direct approach to environmental and cultural concerns by addressing serious ideological, anthropological, and ecological threats.

Three American Indian writers who take activist approaches to environmental concerns in their writing are Simon Ortiz, Wendy Rose, and Gerald Vizenor. Simon Ortiz, in the third volume of poetry from *Woven Stone*, entitled *Fight Back: For the Sake of the People, For the Sake of the Land*, reveals the devastating effects of the uranium industry of the 1960s on the people and the land of the Acoma Pueblo community where Ortiz grew up. In the introduction to *Woven Stone*, Ortiz states that the writings in *Fight Back*, "although not written until twenty years later, as the industry was winding down, were being formed in [his] experience and perception of it in [his] early adulthood" (22–23). It is in this collection that Ortiz provides an historical account of the exploitation of the people and the land along the Grants, New Mexico, mineral belt and most fully voices his ecological, political, and economic concerns for his people and for the American Indian culture as a whole.

Gerald Vizenor's *Bearheart: The Heirship Chronicles* also addresses environmental themes by depicting an apocalyptic wasteland of near ecological destruction. Throughout the novel, the traveling pilgrims come across characters that are victims of the technological violence that has destroyed the Earth's balance. The evil gambler is the embodiment of the greed, selfishness, and environmental destruction that lie at the core of societal corruption. Proud Cedarfair, a trickster and one of the traveling pilgrims, represents American Indian views of nature based on reciprocity and respect, knowing that conquering nature is an illusion. Vizenor also uses this character to confront and undermine static Western culture worldviews and constructions of American Indians by academics and anthropologists. The characters in the novel who hold these "terminal creeds" perish before they make it to the new world.

It is clear that anthropological concerns are important to Vizenor. His *The People Named the Chippewa* is a work of anthropology in and of itself in that it presents narrative histories, memoirs, court testimonies, tribal stories of the trickster, and information by Vizenor on dreaming, naming, the imagination, and language as significant aspects of the Chippewa heritage. In writing this work, Vizenor intended to take back control of the information that has been made available by Euroamerican scholars in order to refute past anthropological inventions of Native cultures.

Wendy Rose, often considered an activist in her poetic writings, presents a direct and sometimes angry voice against the fields of anthropology and archaeology. It disturbs her to know that the bones of her ancestors are excavated, displayed, and sold, turning American Indians into clowns and museum specimens. As part of her activist endeavor, Rose has sought a Ph.D. in the field of anthropology in order to assist the American Indian culture in the proper handling of artifacts. In coming back full circle to American Indian authors at the end of this chapter, I hope to show how various authors are encouraging environmental preservation and cultural change by reterritorializing and recovering their own cultural histories and artifacts.

Finally, the discussions in this book foreground the importance of acknowledging writers who, in the construction of the self, have claimed their mythic, psychic, and environmental terrains and have initiated a movement toward a form of literary decolonization and environmental awareness in which the healing process involves remapping external and internal terrains. Whether by a nostalgic means or imagination as a social practice, retrieval of lost identity and a sense of place is a potent part of the postcolonial industrial world. What many of the writers presented here seem to conclude is that any prior state of cultural consciousness disappeared once colonization occurred, and one can only look back in nostalgia at what was lost and attempt to creatively reconstruct life anew.

Chapter Two

Mythic Reterritorializations

> People say that in the beginning was the word. But they have forgotten the loneliness of God, the yearning for something that shaped itself into the words, *Let there be.*
>
> —Linda Hogan, *Solar Storms*

> Sometimes I see things as they were before this world, in the time of the first people. Not just before the building of houses, the filling in of land, the drying up of water, but long ago, before we had canoes and torches and moved through the wet night like earthbound stars, slow and enchanted in our human orbit, knowing our route because, as Ama said, it had always been our route.
>
> —Linda Hogan, *Power*

In contemporary American Indian literatures, representations of the past are vivid and haunting. Stories and myths from the oral tradition permeate the American Indian voices, combining history with the present to create rich and often complex literatures. I use the word "complex" in the sense that boundaries are blurred between the past and the present, between history and myth, and between the physical and spiritual. It is a literature of at times confrontational anger and forthright resistance as American Indian writers strive to speak for themselves and claim authority over their histories. This activism asserts itself differently in each writer and creates a literature that attempts to heal and decolonize fragmented and disappearing cultures. The stories connect the people to their land, history, and cultural identity. Lee Schweninger agrees in stating that "the earth, the word, the speaker of the word, and the story are inseparable. They exist within the same lines of dependence as the biosphere" (57).

Joy Harjo, N. Scott Momaday, and Linda Hogan present innovative methods of recalling the landscapes of the past to influence those of the present by using tribal stories. Coming from cultures that have historically experienced dramatic psychological, spiritual, and physical dislocations, these writers demonstrate what I will term *mythic reterritorializations* in their works. Mythic reterritorializations take place when writers salvage the stories and places from the past and rewrite them in order to claim an identity and to establish a sense of place concurrent with their present sense of self. I use the term *mythic* here not in the sense of that which is imaginary or false but to refer to stories from the oral tradition that tell of historical events that serve to unfold part of the worldview of a people or to explain a practice or belief. The purpose of this discussion is to investigate how the sense of place is evoked in works by Momaday, Hogan and Harjo and to consider how that sense relates to each writer's sense of self as a contemporary American Indian writer.

N. Scott Momaday: The Great Plains

N. Scott Momaday, in his work entitled *The Way to Rainy Mountain* (1969), links the oral traditions of the past with his sense of the world and his conceptualizations of his own identity. The book resists classification because it challenges Western cultural literary traditions of form and content, which is perhaps part of Momaday's cultural agenda. Writing by authors with displaced histories or identities often "dismantles notions of value, genre, canon, ect. It travels, moves between centers and margins" (Caren Kaplan 359). Written possibly as a memoir or even as an historical document, the book is broken up into sections, one for each page, with each section divided into three parts. The first of the three parts presents a legend or myth from Momaday's tribal heritage. The second part is an historical gloss or factual comment by Momaday, usually in some way related to that legend. The third section is then a personal reflection or observation by Momaday and also is in some way related to the legend in theme or simply a thought or memory of Momaday's evoked by the legend. In structuring the work with three linking voices, Momaday explores his past and links it to the present to reflect on what his tribal history means to him and who he is as a contemporary American Indian writer.

The work is a journey from past to present, from present to past, and from the mythical to the modern. It is a physical journey back to his grandmother's grave and tribal lands, a journey through the imagination, and an internal journey where Momaday comes to a deeper understanding of his own existence. In his article "The Heuristic Powers of Indian Literature: What Native Authorship Does to Mainstream Texts," Kenneth Roemer mentions that the role of the author in *The Way to Rainy Mountain* is highly communal and that Momaday credits his ancestors and family in its creation. In this way, the work also serves as an expression of the Kiowa community and its commitment to survive despite the physical and psychological removals of the culture.

The work begins with the poem by Momaday:

HEADWATERS

Noon in the intermountain plain:
There is a scant telling of the marsh—
A log, hollow and weather-stained,
An insect at the mouth, and moss—
Yet waters rise against the roots,
Stand brimming to the stalks. What moves?
What moves on this archaic force
Was wild and welling at the source. (2)

The final few lines of this poem evoke a great sense of anticipation. Something is coming. Something powerful is stirring near the surface of the earth that is ready to emerge. The poem describes the origins of Momaday's tribe, the Kiowa. The headwaters are the people coming forth where the beginnings of a river strive under yielding roots. There is nothing that can stop this emergence. And so the journey begins.

The prologue and introduction to *Rainy Mountain* are as much a part of the journey as the three movements that follow. Momaday is preparing his readers for what is to come in the rest of the work: the remains of the stories telling Kiowa history and thought and how Momaday makes sense of them. In the prologue, Momaday maintains of the oral tradition, "What remains is fragmentary: mythology, legend, lore, and hearsay—and of course the idea itself, as crucial and complete as it ever was. That is the miracle" (4). For Momaday, *Rainy Mountain* is a journey of affirmation of his culture, his sense of self, and the importance of the imagination and the "idea itself." His reality and his conception of the past are comprised of more than simply what is credited as historical. For him, "the imaginative experience and the historical express equally the traditions of man's reality" (4). Experiencing life and recollecting it involve all the processes of the complex human mind, including the imagination, and Momaday demonstrates this holistic way of thinking in the unique structure of the work.

It is important for non-Native readers to understand the role of the imagination in the Native culture in order to appreciate more fully the literatures. To accept that that which is derived from the imagination and that which has not physically materialized as being significant to one's conception of reality may seem confusing for readers who have been trained to place a strict boundary between fact/fiction or history/imagination. Critic Hayden White, although approaching literary studies from the position of a Western academic, makes a useful argument regarding this boundary and the way that Western readers approach history and the imagination.

In "The Historical Text as a Literary Artifact," White asserts that a historical representation of an event has elements that are inevitably literary and imaginative in nature. When considering how we have been socially constructed, the origins of our identities, we must begin with the past. We begin by tracing back through historical representations to discover how our constructions have developed, which is what Momaday is doing in *Rainy Mountain.* In doing so, White states that the inevitable problem with an objective account of history is that "if the historian is himself a practitioner of it, he is likely to be a devotee of one or another of its sects and hence biased" (395). This theory places a question mark over the Western idea of factual truth in history if it is in essence a set of assumptions that contain as much imaginative input as any "creative" work of literature.

This is not to suggest that there are no facts in history. White argues that "histories gain part of their explanatory effect by their success in making stories out of mere chronicles" and by taking that which is historically fragmentary and applying what he calls "the constructive imagination" to fill in the gaps of the story (397). The historical records, which we call facts, are always insufficient, and historians must fill in the gaps with what they suppose must have happened. In Western cultural notions of history, the historian decides what is most significant by assigning value to particular events and configuring history according to imperatives of one plot structure or another. Thus, such genres as tragedy, comedy, and romance are created (397–398). Once one accepts this inevitable fictionality inherent in the historical text, it becomes more apparent how historical representations of cultures and our lives are then constructions of both imagination and history and form the basis for one's constructed sense of self.

For many American Indian writers, however, this is not a subject of controversy because the imagination has always been a highly valued part of their histories and is not regarded as something less than real. Paula Gunn Allen in *The Sacred Hoop* states that "American Indian thought is essentially mystical and psychic in nature. Its distinguishing characteristic is a kind of magicalness . . . an enduring sense of the fluidity and malleability, or creative flux, of things" (68). American Indian thought also communicates realities in symbolic modes that are tribe-specific so that histories are often self-contained and only fully understood by tribal members. Understanding this way of perceiving reality and how it contrasts to Western notions of history and reality is important when approaching *Rainy Mountain* or any American Indian work of literature.

In "Exploring the Ways to Rainy Mountain," Joan Henley states that "by structuring the narrative into three-part sections, Momaday encourages us to think about the kinds of knowledge we have access to and the authority on which that knowledge is founded" (51). In questioning that knowledge, we become aware of how categories of thought inform our readings and produce the expectations we bring to texts. Momaday's merging of rhetorical modes allows his readers to move beyond those categories and constructions to experience a text where wisdom and expression are results of embracing pluralistic and multidimensional methods of perception. Henley continues: "As the author moves through his narrative, the distinctions between the parts of the sections become blurred, and interdependencies displayed. Personal experience turns back into myth, and the author's own story assumes the didactic function of the traditional tales" (51–52). What emerges from this understanding is an awareness of the complex, multilayered manner in which human beings receive, process, and communicate knowledge.

Momaday begins by describing a plain in Oklahoma where a knoll upon the landscape is an important landmark for the Kiowas. The knoll, "Rainy Mountain," is described as the site of the worst weather and a place of great vast desolation:

> Loneliness is an aspect of the land. All things in the plain are isolate; there is no confusion of objects in the eye, in the early morning, with the sun at your back, is to lose the sense of proportion. Your imagination comes to life, and this, you think, is where Creation was begun. (5)

The journey back to Rainy Mountain to visit his grandmother's grave begins in a place reminiscent of the beginning of creation for Momaday. There is a sense of clarity in this passage as one imagines a desolate plain where objects are clearly visible and identified. It is here where Momaday can most clearly recognize himself amidst the landscape of his people.

The first large section of the book, "The Setting Out," includes the story of the emergence where "the Kiowas came one by one into the world through a hollow log" as well as other stories that demonstrate the tribe's coming into being. In the legend sections, Momaday tells stories of the sun god who, disguised as a redbird, led away a wife and had children with her. He also writes of Grandmother Spider, who advised the twin offspring of the sun, the discovery of Tai-me, who fed the Kiowa when they were in danger of starvation, and the brother who turned into a "water beast" and became the buffalo from which the Kiowa would receive the bulk of their sustenance. This first section tells the stories of the initiation of three symbols for the Kiowa—the sun, the buffalo, and Tai-me. All in one way or another helped assure the survival of the tribe during times of hardship on the landscape of the Great Plains.

The second section, "The Going On," tells stories of continued gain in strength for the Kiowas, but an increased sense of fear and strife begin to become more prominent. Such stories include a man and a woman being attacked by enemies, a buffalo attacking a warrior, and a woman being banished from the village after stealing the meat from a buffalo her blind husband shot with an arrow.

The historical sections indicate the encroachment of such white settlers into the plains as George Catlin and establish the importance of the horse to the tribe. The Kiowa men are described as highly skilled arrow makers and the lives of the women are described as "hard" and that "only the captives, who were slaves, held lower status" (59).

The strength and power of language are particularly apparent in this section in two stories where knowing the Kiowa language saved the lives of the people and helped them to distinguish the presence of an enemy. In one story, a man and his wife are in their tipi working when the man notices someone watching them from the outside of the tipi through a small hole. The man, who was making arrows, addresses the onlooker in the Kiowa language and asks him to identify himself, knowing that if the onlooker were Kiowa, he/she would understand and respond. When the voyeur doesn't respond, the man raises his bow and arrow as if practicing to shoot and aims it around the room. The aim then falls on the place where the enemy is standing, and he is killed by the Kiowa's arrow.

The other story that affirms the power of language tells of the Kiowa making a horse out of clay. When they were finished, the horse behaved violently and caused a tremendous storm. In order to calm the horse, the Kiowa spoke to it in their language. From that day on, the Kiowa know to speak to the storm spirit and say "pass over me" when it comes because it understands their language and will therefore pass.

In these two stories, language is essential in protecting and saving the lives of the people. The first story emphasizes the importance of language in relation to tribal affiliation, and the second story indicates a strong spiritual connection to a higher power through language. Perhaps the significance of language for Momaday is most apparent in a way that is not stated in these stories—for the continuity of tribal histories through the telling of stories. Oral-tradition-based cultures depend foremost on the communication of histories of past occurrences and ancestors through the spoken word for survival. What is perhaps lost in these stories as they are transcribed into the print culture by Momaday is this performative quality, which not only more fully connects the teller with the tale but also adds a dimension of spontaneity as the story changes with the changing recollection of each storyteller. Regardless, the stories do illuminate the intimacy between language and tribal identity, and it is this tribal identity told through myth that Momaday seeks for deeper self-understanding.

The third section, "The Closing In," tells the stories of the final days of the Kiowa people. Momaday, in both the legend and historical sections, details the decimation of the buffalo and the loss of Kiowa tribal power and coherence. A hunting horse turned from his course by a fearful warrior dies in shame; the Tai-me medicine bundle falls to the floor and is stolen, and grandfather Mammadaty loses his temper and accidentally shoots a tame horse as he tries to shoot one that is wild.

The linear structure of Momaday's recollections throughout these three sections indicate the Western cultural and literary influence in *Rainy Mountain*, but this is not unusual for writers of displaced cultures who operate from a liminal or marginal location. What keeps Momaday connected to his origins and to his sense of self is, first of all, the telling of the stories. Momaday believes that it is the idea itself that is the miracle, the very act of saying—this is what has been and this is who we are—is what is most significant: "In the course of that long migration they had come of age as a people. They had conceived a good idea of themselves; they had dared to imagine and determine who they were" (4). And along with these stories that enable Momaday to relocate his idea of himself in his culture are the landscapes and terrains that give them life.

In each of the twenty-four units, the third, italicized paragraph is a personal observation or recollection by Momaday, many of which describe a profound, nostalgic memory evoked by the landscape and life on the plains. In unit two he states: "I remembered once having seen a frightened buck on the run, how the white rosette of its rump seemed to hang for the smallest fraction of time at the top of each frantic bound—like a succession of sunbursts against the purple hills" (12). In other musings, he remembers the quiet and stillness of the plains, swimming in the Washita river, the beauty of horses, the details of a tarantula, the piercing call of a bobwhite, and the thrill of being chased by a buffalo protecting its calf. J. Frank Papovich in his article "Journey into the Wilderness: American Literature and *The Way to Rainy Mountain*," differentiates Momaday's relationship with the Great Plains from the solitary quests and adventures that characterize the way that many protagonists in American literature (usually male) experience place:

> Unlike the view in much of American literature, Momaday's understanding of his place in the world comes not through isolated meditation in a largely inanimate, esthetically distanced environment but through participation in a landscape teeming with profoundly significant life, with horses and buffalo and bear as well as with water and wind and star beings. (123)

The landscape descriptions in the work verify Momaday's intimacy with the landscape and his deep abiding love for it. His active relationship with the landscapes and living creatures of the Great Plains also establishes *The Way to Rainy Mountain* as an important contribution to American literature studies and ecocritical approaches to literary texts.

In unit twenty-one, Momaday recalls his grandfather wondering how it was that the dirt mound around a mole hole was so fine. Mammedaty's question was answered one day when he saw a mole come out of its hole and spit out several mouthfuls of dirt to form the mound. From this occasion, Momaday states, "that was a strange and meaningful thing to see. It meant that Mammedaty had got possession of a powerful medicine" (73). The "powerful medicine" in this case was that of knowledge and blessing. Mammedaty was given the gift of being able to witness this event in the natural world, which few people actually see or experience. This kind of intimate experience with the landscape is honored and emulated by his grandson, who makes the spirit of the plains once again part of his own during this multi-faceted journey to his homeland. This journey for Momaday is "an evocation of three things in particular: a landscape that is incomparable, a time that is gone forever, and the human spirit, which endures" (4). But the journey does not end here. It keeps going in memory, in story, in the imagination, and in the "idea itself."

The Way to Rainy Mountain is a retelling of Kiowa history in order to reestablish and reclaim cultural identity to help ensure cultural survival. It is a restoring of the stories and places from the past and a rewriting of them in order to claim a personal identity for Momaday and to provide a sense of place for him to be most closely connected to his own sense of self. The mythic reterritorialization occurs when Momaday, and we as readers, depart from the text different from who we were when we first began crossing *Rainy Mountain*. Through the stories our territory has changed. For Momaday, a sense of home is reestablished. For his readers, a glimpse into that home entices us to reflect on our own.

Linda Hogan's Terrain of Crossed Beginnings: The Aquatic Territories

Linda Hogan, in *The Book of Medicines* (1993), evokes the oral tradition generously throughout this volume of poetry, but her terrain involves more than the territory of the Great Plains as in *The Way to Rainy Mountain*. It may be inappropriate to refer to Hogan's sense of place as a "terrain" because that term connotes land or physical ground. Hogan's dominant sense of place in this work is aquatic—on the sea, by the sea. She's in the water or the water is in her. Her poems are filled with oceanic images; whales, fish, fisherman, sand, salt, dolphins, ships, and shells are all images that populate Hogan's poetry. In one passage from the poem "Crossings," she traces human origin back to an aquatic beginning:

> Sometimes the longing in me
> comes from when I remember
> the terrain of crossed beginnings
> when whales lived on land
> and we stepped out of water
> to enter our lives in air. (28)

Hogan's desire for water is primal and mythical and is echoed throughout many other poems in *The Book of Medicines*. It is nostalgic for a time when humans had not assumed dominion over the earth but were oceanic creatures crossing paths with the whales in the course of evolution and emergence. According to

Stacy Alaimo, "Crossings" expresses a "yearning to experience bodily ties with nature and to cross over to a time when those corporeal connections were most evident" (59). In addition to highlighting physical connections, the poem suggests an interchange of knowledge. Life that was once aquatic and is now terrestrial (and vice versa) maintains some sense or memory of its prior state of existence and evokes in Hogan the longing for the water that characterizes her sense of origin. Additionally, this "terrain of crossed beginnings" demonstrates the transformative quality of American Indian cultures, adapting to their surroundings in order to survive rather than forcing the environment to adjust to their requirements, a characteristic of Western culture ideologies.

Hogan's poetry is more directly activist than Momaday's work. *The Book of Medicines* is not only a reclaiming of cultural identity and authority over her Chickasaw history but also an ecofeminist endeavor: a voice speaking out against environmental injustice. In an interview with Laura Coltelli, Hogan states, "Spirituality necessitates certain kinds of political action. If you believe that the earth, and all living things, and all the stones are sacred, your responsibility really is to protect those things. I do believe that's our duty, to be custodians of the planet" (*Winged Words* 79). One of the ways in which Linda Hogan acts as custodian is through her writing. In a later chapter, we will take a closer look at her ideas about environmental stewardship and the places we inhabit in her *Dwellings: A Spiritual History of the Living World.* As we explore her poetry in the light of stewardship and myth, we see that Hogan draws attention to injustices committed against her culture, her gender, and the environment. In "Hunger," Hogan personifies physical starvation as a being who "crosses oceans" and "sits on the ship and cries." This evokes the image of colonial settlers crossing the sea in search of that which will fulfill them. These travelers are clearly male:

> Hunger was the fisherman
> who said dolphins are like women,
> we took them from the sea
> and had our way
> with them.

Also:

> It is the old man
> who comes in the night
> to cast a line
> and wait at the luminous shore.
> He knows the sea is pregnant
> with clear fish
> and their shallow pools of eggs. (17)

The language and imagery of these passages are highly sexual, equating an exploitation of natural resources with the exploitation of women. The first passage conjures up images of rape with women as dolphins taken from their aquatic territories and objectified to serve the needs of men. The second passage introduces the sea itself as the figure of a bountiful, pregnant woman who is unaware as the men approach at night waiting to draw from her that which she nurtures.

Criticism of this nature/woman subordination is at the core of ecofeminist theory. Early colonists, before arriving on the new continent, were expecting a Garden of Eden and land of plenty for their consumption. The journals of Columbus describe a lush land with cinnamon and spice, aromatic plants, herbs, flowers, fertile soil, and an abundance of gold. The Spaniards were looking for what they already recognized as resulting in personal or capital gain. This is where the metaphor of nature as virgin woman has its roots.

Annette Kolodny, in *Lay of the Land: Metaphor as Experience and History in American Life and Letters* (1975), examines the land/nature-as-woman symbolization in American literature. She refers to the metaphor as a male fantasy:

> a daily reality of harmony between man and nature based on an experience of the land as essentially feminine—that is, not simply the land as mother, but the land as woman, the total female principle of gratification—enclosing the individual in an environment of receptivity, repose, and painless integral satisfaction. (4)

While "Hunger" takes place on or near the ocean, Kolodny's theories could still be applied here. The poem, situated in an historical context, enables the speaker to tell of the beginnings of colonial settlement that ultimately brought about American Indian cultural genocide and carried with it patriarchal practices enforced by a belief in a divinely ordained mission to dominate and subjugate the "new world." Given that the psychology behind dominating other life forms can also be traced to fear, an aquatic or terrestrial intelligence emanating from the unknown or mystery in the nonhuman natural world would be deemed a threat in need of controlling.

Another example of this patriarchal assault on women and the land is in the poem "Harvesters of Night and Water." This poem begins with the image of men out on a boat in the middle of the ocean fighting to capture a resistant octopus. The boat they are on is described as "white" and "small," and the nets they are using are described as "impotent" and "limp." Again, Hogan uses sexual imagery in depicting the boat and net while no doubt making simultaneous descriptive references to those who occupy the boat. The speaker is sorrowful for the violent and cruel manner in which the men attempt to catch the octopus.

> The tentacles fall down over themselves
> and inch down,
> with the men screaming,
> jabbing at it. I want to stop them.
> I want to tell them what I know,
> that this life collects coins
> like they do
> and builds walls on the floor of the sea. (23)

This passage reveals the frustration Hogan feels in wanting others to respect and value other living creatures. She desires to heal the gap that Western culture has placed between humans and nature by telling the men that the sea creature lives its life to survive as they do. The octopus saves valuables for future use and builds its home underwater in the same way that the men do on land. Hogan is not trying to humanize the octopus in an anthropomorphic sense but rather to foster a respect: this creature is a living being and not simply an object for their capture and consumption. Furthermore, the objectification of nature that allows these men to brutally attack the octopus is a result of the notion that human beings are not part of nature. Otherwise, there would be an understanding that the men are brutalizing a part of themselves. Hogan ends the poem with this stanza:

> I want the world to be kinder.
> I am a woman.
> I am afraid.
> I saw a star once, falling toward me.
> It was red
> with brilliant arms
> and then it was gone. (24)

Hogan's call for a kinder world is echoed by the vanishing falling red star, which may symbolize the strength and brilliance of the Native cultures. As a woman, she is aware there is much to be afraid of. The harvesters of the night can kill and exploit women in the same manner that they brutally attack the octopus. This disrespect for other life forms has created a cruel and brutal world, in Hogan's view. The men who are harvesting the night in the seas that she navigates in her poetry fail to acknowledge and respect the aquatic intelligence that surrounds them.

Hogan's natural world in *The Book of Medicines* is beautiful, wild, and also very dangerous. It is not nature that is depicted as dangerous, but rather the colonizers who attempt to abuse and destroy the land and her people. In "Tear," for example, Hogan gives the title word double meaning in order to describe the dresses the women in her tribe wore and the historical removal that devastated her ancestors on the western pathway known as the Trail of Tears.

> Tear dresses they were called
> because settler cotton was torn
> in straight lines
> like the roads we had to follow
> to Oklahoma.
>
> But when the cloth was torn,
> it was like tears,
> impossible to hold back,
> and so they were called
> by this other name,
> for our weeping. (59)

The settler cotton torn in straight lines, like the straight roads to Oklahoma and the streaming tears running down the faces of the people forced to march this harrowing path, is the cloth of the colonizer. The term "tear" in its one definition suggests a rupture or split and reminds Hogan's readers of the separations her ancestors underwent during the events surrounding the Trail of Tears. Separations from family members, homes, tribal regions, from their ways of life, their languages and identities are just a few of the various forms of human and cultural fractures that are a result of the colonial process.

In *The Book of Medicines*, the poems with water imagery are compelling in expressing Hogan's equation of the subjugation of women with the exploitation of nature and American Indian cultures. While in "Tear" the water imagery navigates readers through the flood of sorrow caused by the history of cultural displacements, much of the time the water imagery signifies birth, hope, cleansing, unpolluted earth, and healing. The aquatic medicine in Hogan's work has many different healing purposes. The healing comes about through loving, in remembering the past, in telling the stories, in writing the words, and in respecting the earth and the wisdom that it provides. The final poem in *The Book of Medicines*, "Flood: The Sheltering Tree," contains the image of Hogan standing beneath the only tree existing on a mound where water keeps slowly rising as "Land takes back the forgotten name of rain / and speaks it / like a roar, dark and running / away from breaking sky" (85). This rain is the tears of her people washing the wound and cleansing the land so that they can see again "the beautiful unwinding field / and remember [their] lives / from before the time of science, / before [they] fell from history" (85). There is a nostalgic mood evoked with a longing for a time when the Chickasaw culture was alive and thriving. This was a time when the earth still had wide-open spaces, before the European settlers brought their philosophies.

In *The Book of Medicines*, Hogan has reterritorialized a space that is nostalgic in its mythic reminiscings and honest in its portrayals of environmental injustice and sexism. Hogan has created a spiritual space with water as the primary ingredient needed to heal and reconnect her with her cultural heritage. Ironically, water is also the primary ingredient she uses to describe portrayals of environmental injustice, racism and sexism. The aquatic intelligence permeating her poems emphasizes the spiritual perspectives from which Hogan writes. In recognizing that intelligence, Hogan accepts the responsibility that accompanies it—that we all need to behave as custodians of the planet and work harder to protect the earth and its inhabitants, terrestrial or aquatic.

Joy Harjo: The City of Terrible Paradox

In an interview with Donelle R. Ruwe from *Joy Harjo: The Spiral of Memory: Interviews*, Joy Harjo states, "I believe myth is an alive, interactive event that is present in the everyday" (Coltelli 130). For Harjo, descriptions of landscapes and settings from the oral tradition are combined with modern cityscapes, often creating a surrealistic and sometimes disjunctive plane of existence. Although Harjo's poetry can seem complex, her poems exemplify American Indian beliefs that embrace transcendence and transformation—of time, of place, and identity. Harjo is able to juxtapose cosmic and mythic aspects of her culture with mundane objects of the present world, creating an atmosphere that finds the spirituality in both. According to John Scarry, "Harjo's simultaneous physicality and spirituality, and her ability to combine the eternal past and the continuing present . . . are among the most noteworthy of her characteristics" (287). These characteristics in Harjo's poetry culminate in a writing style that incorporates elements of both her Native culture and the Western culture in order to produce a unique voice with a mystical and adaptive sense of place.

In the essay, "Towards a National Indian Literature: Cultural Authenticity in Nationalism," Simon Ortiz discusses the "creative ability of Indian people to gather in many forms of the socio-political colonizing force which beset them and to make these forms meaningful in their own terms" as a means of cultural resistance and part of the struggle for liberation (8). The poetry of Joy Harjo highlights this creative ability and resistance in the manner in which she manipulates poetic conventions and ideas of time and place to create a world distinctly her own.

The title poem from *The Woman Who Fell from the Sky*, (1994), brings the ancient legend into contemporary times by telling the love story of a woman who fell from the sky toward a world of concrete and tin cans in order to be caught by Johnny, a war veteran. The oral tradition story about the woman who fell from the sky is a tale of origin and creation that cannot be adequately summarized because each tribe has its own version of the story. In a very general sense, it entails a woman slowly falling from an upper world for many years toward a land that has been flooded. Several animals survive the flood, a turtle and a beaver most prominently. The beaver dives to the bottom of the ocean, collects mud, and places it on the turtle's back. The mud becomes larger, forming the North American continent, or Turtle Island, where the woman who was falling then safely lands and begins the human race. Harjo's version of the story is one adapted for a modern world complete with asphalt, a Dairy Queen, and flying trash.

The poem takes place in a "city of terrible paradox," where an "ordinary" woman, Lila, is falling from the sky toward Johnny, named Saint Coincidence. In this poem, Harjo employs a poetic contrapuntal technique that interweaves two stories so fluidly that one can barely extract one from the other. There is on one hand, the story of Lila and Johnny, who met at an Indian boarding school several years earlier but separated when Johnny left to join the army. The other story is the mythical story of the woman who fell from the sky after climbing up a stairway to the stars: "She dared to look back and fell. Fell through the centuries, / through the beauty of the night sky" (8). Throughout the poem, Harjo weaves the falling woman story of Lila back and forth between the mythical and physical worlds, narrowing the space between the material and spiritual.

After graduating from Indian boarding school, Lila worked cleaning houses and at a Dairy Queen "wearing her uniform spotted with sweets and milk." The poem concludes with Harjo returning Lila and Johnny to mythical figures by presenting the final uniting of the two lovers within the context of the story from the oral tradition but within a remarkably modern environment: "She fell and was falling when Saint Coincidence caught her in / his arms in front of the Safeway as he made a turn from borrowing / spare change from strangers" (9). The imagery indicated in this poem, and many others by Harjo, suggests that Harjo's home and sense of place are not bound by physical limitations but rather transcend and travel into the spiritual and back again at the poet's will.

She writes this passage in an italicized section after the poem: "I traveled far above the earth for a different perspective. It is possible to travel this way without the complications of NASA. This beloved planet we call home was covered with an elastic web of light." And from this different perspective, Harjo saw "revolutions, droughts, famines and the births of new nations. The most humble kindnesses made the brightest lights. Nothing was wasted" (10). Similar to the structure Momaday uses in *The Way to Rainy Mountain*, Harjo poems are broken into two parts separated by a star. The second part is a personal thought or memory of traveling beyond the earth's sphere evoked by the previous poem. From that unique perspective, Harjo "understood love to be the very gravity holding each leaf, each cell, this earthy star together" (10). This final statement reflecting on the poem clarifies for the reader that love is the ultimate theme. Through love, the mythical and the modern can coexist, and all that inhabits the earth can be held together.

Like Momaday, Harjo has retold aspects of the creation myth, reconfiguring a story for the modern world and landscape that she now inhabits. This mythic re-creation of territory alters the story and bridges the gap between the past and present, the spiritual and mundane, history and the imagination. The new poem that is created is a tale of longing and suffering where transformation is necessary for survival. The territory is stark and beautiful at the same time as the supernatural world bleeds onto the cityscape. This reterritorialization mirrors an attempt to maintain a connection with one's heritage by embracing cultural beliefs and traditions within a modern territory of "terrible paradox." For Harjo, the stories and the spirit provide that connection.

Harjo demonstrates the same technique in "Deer Dancer" from *In Mad Love and War* (1990), by having a mythical woman appear in a bar and begin dancing in front of the "Indian ruins" who are drinking beer and playing music on a jukebox. The mythical woman is a blessing, a connection to the heritage and past that keeps them going. The way Harjo brings the mythic into the modern and combines the spiritual with the mundane provides readers with a poetic voice emerging from a postcolonial legacy. This use of the mythical deer woman figure serves as a means to relocate the dislocated narrator back to the spiritual and natural worlds of her ancestors. In the italicized section of "The Postcolonial Tale," from *The Woman Who Fell from the Sky*, Harjo writes, "The landscape of the twentieth century is littered with bodies of our relatives. Native peoples in this country were 100 percent of the population a few hundred years ago. We are now one half of 1 percent. Violence is a prevalent theme in the history of this land" (19). Violence, war, alcoholism, and suffering are also part of the history and present condition for Harjo's people as described in "Deer Dancer." She describes the bar as "the bar of broken survivors, the club of shotgun, knife wound, of / poison by culture" (5). At the end of the poem, however, the mythical woman evokes images of peace and beauty and life. The landscape of her ancestors becomes transformed from a barroom filled with the beaten and lost human shadows of a time long past to a forest at dawn where there is no hunger, and her people and her community remain unharmed.

Additionally, music is a thematic and driving undercurrent in Harjo's work that connects her to the landscapes of her ancestors and plays a role in forming her poetic voice. She has been highly influenced by the visual arts, filmmaking, and music, with music being the most dominant influence. "She has been described as listening to music more than reading the work of other poets, and Harjo herself has said that when she writes poetry she does not start with an image but rather with a sound" (Scarry 286). A saxophonist and a member of a jazz band, Harjo has a clear understanding of jazz techniques, and she is able to apply them to her artistic efforts, such as in the contrapuntal technique she uses to weave stories in her poetry. She has released audio recordings of several of her jazz band performances and often plays the saxophone during her poetry readings. It would perhaps be an understatement to assert that Harjo recognizes the relationship between jazz and her poetry as unified and constant and that the two artistic mediums enhance each other and work together to form her uncommon expressive voice. Equally important is the role music plays in providing Harjo with images and metaphors that clarify and heal Native confrontations with Western philosophies.

In "Original Memory," Harjo contemplates the importance of music for her people as a element that helps define and clarify events. Through music, the gaps that exist between past and present due to the linear Western concept of time can be filled as the music evokes recollection of events from the past and allows her people to embrace and let go of a bitter history. In her poetic passages, music is a transformer that reshapes linear time into a cycle where the space between past and present folds and provides access to events and ancestors from another time. Her poetry and music not only help to bridge the gap between the past and the present for American Indian cultures whose foundations and beliefs depend on stories and memories, but they also help to heal the wounds from the physical and spiritual displacements which have made Native writers feel dislocated from themselves, their homes, and their heritages.

Stories, poetry, memories, and music bring Harjo in contact with her cultural heritage by evoking memories of the places that were an integral part of the events and traditions of her ancestry. A sense of home cannot be conceived in isolation from the landscape that paints its background. The concept of "home" for Native women activists, such as Harjo, is the subject of an article entitled "Relocations upon Relocations: Home, Language, and Native American Women's Writings," by Inés Hernández-Ávila. "Relocations" states that "the concern with 'home' involves a concern with 'home*land*.' Even when Native women activists no longer reside on their ancestral land bases (though many still do) they continue to defend the tribal sovereignty of their communities as well as communities of other indigenous peoples" (492). This defense of tribal community is particularly evident in Harjo's poetry in the way that she travels from one landscape to another, from one sphere to another, but so often returns in her poetry to Oklahoma where her tribal heritage has its origins. Much of her earlier poetry is predominantly situated in and about Oklahoma, telling of the Creek ancestors and descendants who remain, while her more recent poems portray Harjo in many different landscapes, cityscapes, hemispheres, and worlds. Harjo clarifies this for her readers in an interview with Sharon Stever from *The Spiral of Memory*. She states:

> my overall sense of home means something larger than any place nameable here in this land; it's as if this land is of that larger place, a hint as to the larger story, and it makes a spiral. The poem then becomes a home, sometimes with a glimpse, an eye toward the story of origin, or a place for the human understanding of a hummingbird. (Coltelli 76)

If poetry is a home for Harjo, then one can theorize that Harjo's home, among other things, is always moving, traveling, weaving, transforming, and transcending the limitations of physical bounds. In "Transformations" from *In Mad Love and War*, Harjo most clearly depicts this point by transforming and inverting the meaning of hatred and the meaning of poetry. She begins the poem by writing, "This poem is a letter to tell you that I have smelled the hatred you have tried / to find me with; you would like to destroy me" (59). In the middle of the poem, Harjo shifts from a mystical voice to one that is lucid by directly reflecting on the poem and pointing out the transformation that is occurring within its composition.

Harjo writes that it is possible to turn a poem and hatred into a "blackbird, laughing" or a "piece of seaweed" through language and meanings (59). Her captivating images from nature remind us not only of the power of language but also of the power we have as human beings to redefine and alter aspects of our lives that may cause discomfort. This reconfiguring at one's will of the place, circumstances, and meanings that construct an existence is at the heart of reterritorialization. By the end of the poem, the hatred initiated in the first few lines of the poem has been inverted and transformed into love.

In an autobiographical essay entitled "Ordinary Spirit," Harjo discusses how this poem is also about processes, the process of change, "the process of becoming," and the process of how a poem becomes a poem: "Within the poem is the process of the 'hater' becoming one who is loved, and who ultimately loves" (269). Through transformation, and "the process of becoming," the poem reflects the need to question conventional ways of viewing the world and oneself in order to evolve one's own identity.

In addition to this transformation of the human being, the poem also transforms the notion of poetry. By having the poem be able to change into animals located in the natural world and ancient myth—a bear, piece of seaweed, or a blackbird—Harjo's poem, and home, becomes a landscape littered with animals and love rather than with violence and the dead bodies of her ancestors. Like Momaday, Harjo alludes to the importance of the "idea itself," the magic that can happen when human beings are willing to use their imagination and conceive something into being.

For Harjo, Momaday, and Hogan, home is a story not only of the past but a story of now. The identities they construct through their literature and stories would not be complete without the accounts of colonization and suffering that have led to the internal and external struggles they continually face today, such as the loss of the self and racism. Their writing demonstrates "what Chela Sandoval calls 'oppositional consciousness,' the ability to read and write culture on multiple levels" (Caren Kaplan 357). They write from the perspectives of both cultural insider and outsider, using the colonizer's language, which enables them to both conform and resist in the struggle for cultural identity and liberation. Mythic reterritorialization is a form of resistance where the stories from the oral tradition have been remembered, rewritten, and relocated by writers who choose to identify themselves and their histories on their own terms regardless of what conventions they break or what symbols are not understood by Western cultures. In "Towards a National Indian Literature: Cultural Authenticity in Nationalism," Simon Ortiz asserts that,

> the continued use of the oral tradition today is evidence that the resistance is still on-going. Its use, in fact, is what has given rise to the surge of literature created by contemporary Indian authors. And it is this literature, based upon continuing resistance, which has given a particularly nationalistic character to the Native American voice. (10)

Additionally, it is not only the continued use of the oral tradition that acts as a form of postcolonial resistance by American Indian writers but also the retelling and relocating of the mythical stories and histories with their own voices, on their own terms, and in their own places.

The three authors studied in this chapter, Joy Harjo, N. Scott Momaday, and Linda Hogan, reterritorialize the landscapes of the past to influence those of the present using myth and the oral tradition. In their literature is the quest to tell the stories of place through writing about the plains, the oceans, and the surrealism of Harjo's cities, to make us all conceive of these places in new ways. Their stories and territories leave nothing out. They are filled with the beauty of their myths, the tragedies of the American Indian holocaust, and the transforming powers of hope, love, and the imagination.

Chapter Three

Psychic Reconfigurations of Culture and Place: Sites of Confrontation and Refuge

> There was a sound of heavy dropping of rain from the eaves, and the distant roar and undertone of the sea. My thoughts flew back to the lonely woman on her outer island; what separation from humankind she must have felt, what terror and sadness, even in a summer storm like this.
>
> —Sarah Orne Jewett, *The Country of the Pointed Firs*

> Every increment of consciousness, every step forward is a *travesía*, a crossing. I am again an alien in new territory. And again, and again. But if I escape conscious awareness, escape "knowing," I won't be moving. Knowledge makes me more aware, it makes me more conscious.
>
> —Gloria Anzaldúa, *Borderlands/La Frontera*

Remapping the terrains, whether in a metaphysical or geographic sense, is always to some degree psychological. The desire for that inexplicable feeling of home starts with a vision, a memory, or a nostalgic yearning involving all the senses envisioning and/or recollecting the details of place within the mind. As the psyche constructs the territory, choices are made, not only of where but also of who and what stays or goes. Home not only encompasses the sense of place but also an environment in which people feel accepted and loved for who they are and where they are free from oppressive forces.

For authors Chrystos, Gloria Anzaldúa, and Susan Griffin, conflicts stir and confrontations take place across the hazardous terrains where they walk. These writers have made it their goal to unapologetically redefine the ground they traverse and change the consciousness of their readers through writing. This activism manifests itself by directly confronting issues of sexism, racial oppression, environmental degradation, and border culture struggles on each author's chosen psychic territory. Chrystos, Anzaldúa, and Griffin have each undergone some form of deterritorialization by oppressive, patriarchal structures, which have defamiliarized them from their notions of self and access to basic human rights. The deterritorializations and the psychic reterritorializations are present in the form of landscapes or places used literally and metaphorically to represent sites of conflict or refuge where the writers criticize and reconstruct borders and zones of human and ideological contact.

Chrystos: The Battlefield and Floral Terrain

In the final poem of her fifth volume of poetry, entitled *Fire Power* (1995), American Indian poet Chrystos writes:

> I consciously fashion my writing as a weapon for my own survival & that of my allies. It has surprised me that my objective has so often been overlooked by reviewers, who consistently reduce me to "feeling angry," when my purpose is so blatantly to shoot back in the old way, from the trees, in sneak attacks that masquerade as strings of artfully arranged words. (128–129)

The poetry of Chrystos, offering a voice that projects more than simple verbalistic venom, speaks from a culture whose voices have been throttled by Western cultural impositions. With the stifled voices of her ancestors speaking through her, is it any wonder that she emerges from her poetry with explosion? The power and unapologetic stance of many of her poems, particularly in *Not Vanishing* (1988), may account for the reductionist position reviewers often take when writing on her works. As Chrystos mentions above, however, her objective goes beyond mere confrontational outrage. She writes, "I am a warrior against all forms of injustice," and her poetic text forms the battle zone from which she fights and finds her means of survival (*Fire Power* 129).

Although Chrystos does hail herself as a warrior, her technique in battle is a carefully crafted artistic voice. And within the rough yet lush and beautiful terrain of her textual battlefield, Chrystos survives by taking control of her own life, cultural heritage, and community. Emerging from a historical legacy characterized by forced assimilation, issues of identity are crucial in American Indian literary voices. According to Sidner Larson, "it is well-recognized that the original American Indian notion of identity was communal, or tribal. With colonization, however, came extreme pressure to abandon the tribal notion of identity in favor of individuality, a divide and conquer strategy" (57). In *Not Vanishing*, Chrystos's counterattack to this strategy involves constructing a community where she can survive and heal. She achieves her aim by blatantly rejecting particular characters in her poems and embracing others. In accordance with Bakhtin's theory of dialogics, Chrystos presents a social context in her works that enables her to dialogue with individuals, social groups, and ideologies which have a place on her battlefield, whether as zones of hostility or as sites of refuge. This dialogue plays an essential role in the survival of Native cultures and for Chrystos herself.

The relationship between the love poems and the polemical poems in *Not Vanishing* is nothing less than dichotomous. They are not separated in the book, but rather interspersed, forcing the reader to rethink constantly his/her own position in the text. "'Finding oneself on the outside,' extralocality, is, therefore, a determining condition of the literary word, in the same way as is participation in life, in the contents and values of social life," according to Augusto Ponzio (4). Chrystos deliberately establishes a threatening relationship with particular members of her audience and characters in her poems, knowing that survival on her battlefield entails destroying that which has armed oppression.

Poems such as "White Girl Don't" and "I Am Not Your Princess" harshly address specific white audiences who participate in her textual, external dialogue by providing a locus of resistance. "White Girl Don't" pronounces disgust toward the "white girl" who claims knowledge of cultural pain and suffering from her privileged position from a Western cultural background. Chrystos writes: "We aren't the latest fad in your candy-striper life / You want genocide / look out the window at the road going past your house / honey / it's killing us" (75). This poem amplifies the connotations of the book's title by addressing the girl who talks about tragic conditions in other countries but is unaware of the suffering in her own. The white girl, blind to U.S. poverty, violence, and cultural genocide, evades her own answerable consciousness: "Somewhere else is safer & not your [her] fault & not your [her] responsibility" (75).

The poem, exposing the white girl's privileged class status and cultural ignorance, reveals how the white girl participates in her home country's oppression by failing to recognize its existence. According to M. M. Bakhtin,

> what underlies the unity of an answerable consciousness is not a principle as a starting point, but the fact of an actual acknowledgment of one's own participation in unitary Being-as-event, and this fact cannot be adequately expressed in theoretical terms, but can only be described and participatively experienced. (40)

In the poem, Chrystos forces the white girl into a participative dialogue in order to inform her of the genocide that takes place right outside her front door. Under the assumption that the American Indian holocaust exists as a static, historicized event, the white girl is insensible to the ways in which the genocide has resonated in her own life, making her responsible for its existence as an ongoing event. Chrystos ends the poem with this assertion to her:

> Don't aim 5,000 miles away to a land whose words
> you barely speak   if at all
> Right here now   genocide
> I'll tell you about it   (75)

The poem "I Am Not Your Princess," addressed to white feminists, expresses resistance against those who attempt to use Chrystos or other American Indians as a bridge between the two cultures to ease their own internal sense of guilt. In this poem, Chrystos locks these women out of her community knowing that she does not owe them anything: "I won't chant for you / I admit no spirituality to you / I will not sweat with you or ease your guilt with fine turtle tales" (66). Chrystos knows that she cannot fight her battle alone, that strength is most powerful when its resistance is collective, but waging a war against an ideology as heated as racism cannot occur without painful confrontation: "Paradoxically, killing is sometimes the only means of preserving life" (DeShazer 275). In order to survive, Chrystos uses her poetic text to publicly denounce those who have inhibited liberation. "I Am Not Your Princess" rejects those who bring their gross generalizations and assumptions about American Indians to Chrystos and expect her to fulfill them.

Equally combative in its dialogue with an external ideology is the poem entitled "Maybe We Should Not Meet If There Are No Third World Women Here." This poem powerfully expresses Chrystos's fury at the apparent invisibility of women of color at a large feminist gathering dominated by whites. In the beginning of the poem she asks: "How can you miss our brown & golden / in this sea of pink" (13). But they do miss it. According to Chrystos, racism and homophobia within the women's movement are pervasive and unconscious. "All those workshops on racism won't help you open your eyes & see / how you don't even see us." An article by Chandra Talpade Mohanty entitled "Under Western Eyes: Feminist Scholarship and Colonial Discourse," a discussion of the reality of racism within the women's movement against Third World women, uncovers the bipolar relationship between the way Western feminists view themselves and the way they perceive Third World women. This Third World woman image presents a woman whose life is:

> based on her feminine gender (read: sexually constrained) and being "third world" (read: ignorant, poor, uneducated, tradition-bound, religious, domesticated, family oriented, victimized, etc.) (199–200)

This perception, Mohanty suggests, is in contrast to the self-defined Western woman who perceives herself as educated, modern, in control of her own body and sexuality, and free to make her own choices regarding her life and future (199–200).

This Western construct of the Third World woman implies that these women are still unempowered victims existing under patriarchal domination and subscribing to fixed gender roles that predetermine and stifle their individual growth. Racist in itself, this power paradigm objectifies Third World women into broad categorizations of powerlessness and subordination. Ultimately, the images that have been created to represent women of color "exist in universal ahistorical splendour, setting in motion a colonialist discourse which exercises a very specific power in defining, coding, and maintaining existing first/third world connections" (Mohanty 214). It essentially sets up the same kind of power paradigm that men have created toward women in order to maintain a dominant position.

Echoing the concerns expressed in Mohanty's article, Chrystos writes: "Don't write a paper on it for me to read or hold a meeting in / which you discuss what to do to get us to come to your time & place / We're not your problems to understand and trivialize / we don't line up in your filing cabinets under 'R' for rights" (13). This passage supports Mohanty's claim that white feminists view Third World women as "a category of analysis" and "as a homogenous 'powerless' group often located as implicit *victims* of particular cultural and socio-economic systems" (200). Getting Third World women to attend a feminist gathering is problematic in that if Western feminists assume Third World women to be oppressed and less liberated, their goal may be to assimilate women of color into their superior image construct. If a woman is domestic, traditional, and religious, does that make her oppressed? By Western standards, yes it does. But, perhaps there are many different kinds of feminism and many ways to be a strong, competent woman. Or, does feminism mean having the choice?

Is it time for Western feminism to redefine itself and the way it perceives women of color? Mohanty calls for a historical and cultural specificity that reduces assumptions, generalizations, and categories that universalize for a more human approach to racism. For Chrystos, the battle is more complicated. She asks, "How can we come to your meetings if we are invisible" (13). Paradoxically, this poem exposes how Western feminism, an organization founded on the empowerment of the oppressed, participates in the suppression of a woman of color. The dialogue Chrystos undertakes with Western feminists in this poem indicates a rejection of their membership in her community.

At this point we have to ask, where, if any, are the safe territories in this textual battle zone where strength and community can be found? Can we say that Chrystos has found peace and sisterhood within a community of Third World women and lesbians? Gloria Anzaldúa in *Borderlands/La Frontera* (1987) states that "it is not enough to stand on the opposite river bank, shouting questions, challenging patriarchal, white conventions. A counterstance locks one into a duel of oppressor and oppressed; locked in mortal combat, like the cop and the criminal, both are reduced to a common denominator of violence" (78). Anzaldúa then suggests the option between either leaving the opposite river bank or crossing "the border into a wholly new and separate territory" (79). The stark contrast between the love poems in *Not Vanishing* and the poems of resistance suggests that Chrystos can cross into a new territory where she disengages from her battle for moments of life-affirming beauty.

Primarily, the erotic poetry in the text employs language that is highly sensual and imagistic with a centered, symmetrical presentation on the page suggesting a sense of order not always discernible in the other poems. The landscape of these poems is rich in vegetation and colorful floral imagery, such as in the poem "Meditation for Gloria Anzaldúa" and "O Honeysuckle Woman." For Chrystos, they provide a site of refuge, a safe community, on her battlefield, where she can exist within an atmosphere that is loving and pleasurable.

In the poem entitled "Foolish," Chrystos, identifies a community from which to speak: *"We're beginning!* First time arrives with yellow smells / surprises These friends I planted rise up to embrace me / All the people are buds"(3). Situated within a lesbian community, Chrystos echoes the empowerment portrayed in Audre Lorde's poetry where "sisters who lay in one another's arms may also bear arms together one day, stronger for having shared the erotic experience" (DeShazer 268). Erotic and sensuous language, therefore, becomes not only a source of power but also an opportunity of sharing among women, which ensures they're stronger when once again they are called into battle. The nature imagery of "Foolish" positively suggests the "Spring renewal of plants, implying a certain inevitability to the envisioned return of native values" (Murphy 45). The people, as living plants and referred to as "buds," are hopeful, which counteracts many of the other poems in the volume which describe genocide and contemporary American Indian struggles, such as racism, poverty, unemployment, alcoholism, and abuse.

Chrystos finds reassurance from the American Indian community in poems such as "I Walk in the History of My People" and "Vision : Bundle," which reaffirm cultural continuity in the face of adversity. Both of these poems demonstrate the distinct cultural situation that has formed her artistic voice. Bakhtin states that "every utterance is the product of the interaction between speakers and the product of the broader context of the whole complex social situation in which the utterance emerges" (Bakhtin, *Freudianism* 79). "I Walk in the History of My People" places

Chrystos within the historical context of her ancestors with references to the Battle at Wounded Knee and the walking images alluding to the Trail of Tears. Although Chrystos was not present at these historical events, she participates in them as ongoing events by embracing her ancestors' suffering within her present physical being. She writes, "In my marrow are hungry faces / who live on land the whites don't want" and "My knee is so badly wounded no one will look at it / The pus of the past oozes from every pore" (7). Chrystos feels answerable to these people who live in her joints and bones, and she takes on the responsibility of continuing to fight the war for those who are no longer able. These poems, these acts of defiance and resistance against oppression emerge from this history and from these ancestors, among whom she identifies herself. "From within the answerable act, the one who answerably performs the act knows a clear and distinct light, in which [she]he actually orients [her]himself" (Bakhtin 30).

From this cultural orientation, Chrystos uses intonation carefully as a poetic device to aid in her survival. The anger that she expresses in this poem has an important function. On the surface, it describes the pain she feels toward the atrocities placed upon her people, but on another level it situates her within a specific community, which is crucial to her survival. Intonation, anger in this case, "requires the choral support of surrounding persons" and inherently involves "a living, forceful relation with the external world and with the social milieu—enemies, friends, allies" (Bakhtin, *Freudianism* 104). With her anger, Chrystos takes an active role in positioning herself in a confrontational dialogue with the forces that have oppressed Native cultures:

> My knee is so badly wounded that I limp constantly
> Anger is my crutch   I hold myself upright with it
> My knee is wounded
> see
> How I Am Still walking (7)

On her battlefield, it is this anger and the community from which it emerges that keep her fighting despite her many wounds.

"Vision : Bundle" uses the same kind of intonation tactic and further situates her within her Native community by referring to herself and other Natives as "we" and the white culture as "they" or "them." Although this poem sets up the oppositional territories which Anzaldúa suggests at some point must be relinquished, Chrystos takes the opportunity to describe a profound way in which the culture is continuously exploited. The poem "describes the destruction of Native spiritual practices by imperialistic anthropology, museum displays, and unauthorized ethnographic recordings of religious rituals. These violations are linked to a reliance on technology and the resulting evisceration of the human spirit" (Murphy 45). The harrowing paradox suggested by this situation is the desire of the imperialistic culture to unearth and preserve that which it has nearly destroyed.

It appears to be acceptable for Native cultures to exist as long as they are frozen within glass display cases, where they have no power. Ongoing disputes continue over ownership of Native materials and skeletons that are excavated by builders as development continues to encroach into ancient Native communities and sacred burial grounds. Like "I Walk in the History of My People," "Vision : Bundle" ends with a proclamation of survival: "The only part of us they can't steal / is what we know" (21). Chrystos knows that the consciousness of her people and the beliefs that they hold to be true can never be stolen or reduced to exhibition. Within the realm of this knowledge exists a safe refuge on her battlefield where the enemy can never enter.

As a "warrior" fighting against injustice, Chrystos raises her weapon to a multitude of corruptions in *Not Vanishing.* Mary K. DeShazer, in *Writing the Woman Artist: Essays on Poetics, Politics, and Portraiture,* contends that "many women of color do claim a warrior identity, especially in their poetry—an identity re-visioned not as a necrophilic zest for destruction but as an ongoing commitment to radical change" (265). By setting up her text as a battle zone situated within a dialogical, social context, one can begin to recognize who and what comprise her allied forces and which remain adversarial. Survival depends on this dialogue, which provides communal support from lovers and ancestors while confronting Western powers to raise consciousness of American Indian histories and current concerns. If colonial forces ensured themselves conquest through a "divide and conquer" strategy, the poetic text of Chrystos regroups and retaliates to survive and heal.

Gloria Anzaldúa: The New Border Consciousness

Politically and personally motivated, Gloria Anzaldúa reconstructs her identity and her psychic terrain in *Borderlands/La Frontera* (1987) in order to re-create a cultural identity compatible with her sense of self. In this work, she discusses the difficulties of border life and how being of Mexican, Indian, American descent, along with being a woman and a lesbian, leaves her open to rejection from all sides of her heritages. She is rejected from her Mexican heritage for being lesbian and from her white heritage for being a woman of color.

Anzaldúa also is the one doing the rejecting of certain aspects of her heritages. She rejects her Mexican heritage when she states "I abhor some of my culture's ways, how it cripples its women, *como burras*, our strengths used against us, lowly *burras* bearing humility and dignity. The ability to serve, claim males, is our highest virtue. I abhor how my culture makes *macho* caricatures of men" (21). While she is criticizing the Mexican culture for its gender constructions that serve to separate men and women and keep women subjugated, she also criticizes American culture for its colonization and exploitation of the land and indigenous and Mexican peoples: "In the 1930s, after Anglo agribusiness corporations cheated the small Chicano landowners of their land, the corporations hired gangs of *mexicanos* to pull out the brush, chaparral and cactus and to irrigate the desert" (9). The colonization involved human and territorial appropriation by the whites that used the Mexicans against themselves, paying them to participate in their own exploitation and landscape destruction.

Anzaldúa throughout the text writes in poetry and prose about the border landscape. The first poem of the text describes the border at the ocean where the waves are "gashing a hole under the border fence" (1). The borderlands are geopolitically represented by the border, which separates the United States from Mexico. In the beginning of the poem, she describes this geopolitical border metaphorically as a physical border across her body. It is a

> 1,950 mile open wound
> dividing a *pueblo*, a culture
> running down the length of my body,
> staking fence rods in my flesh,

splits me splits me
me raja me raja
This is my home
this thin edge of
barbwire. (2–3)

The fact that Anzaldúa describes her "home" in terms that indicate a violent, painful rupture is significant for her in describing the difficult psychological fragmenting that occurs when being torn by conflicting cultural codes. The physical borderlands, "where the Third World grates against the first and bleeds" (3), become a psychological fault line where plates constantly slip and cause confusion and destruction for those who inhabit this land.

The end of the poem heralds the sea and its disregard for the boundaries human beings place on the landscape, while also making a final claim that the landscape belonged to Mexico once and will be retrieved:

But the skin of the earth is seamless.
The sea cannot be fenced,
el mar does not stop at borders.
To show the white man what she thought of his
arrogance,
Yemaya blew that wire fence down.

This land was Mexican once,
was Indian always
and is.
And will be again. (3)

Images throughout the poem connect Anzaldúa to the sea. The sea is a place where she feels peace and refuge from the Western culturally imposed border, the open wound that splits her body. Healing can begin to take place as she internalizes the seamless skin of the earth and, in writing the text, removes the psychic borders that have split her identity behind fences, barbed wire, and steel curtains.

In order for her to locate a safe psychic territory, she creates what is called the New Mestiza, which resists monoculturality and is a blend of all three heritages and the aspects therein that she chooses to accept. This new Mestiza consciousness is Anzaldúa's way of reterritorializing her own psychic configurations of culture in order to live in peace with her sense of self and with the world around her. In the preface to *Borderlands/La Frontera,* she states:

> This book, then, speaks of my existence. My preoccupations with the inner life of the Self, and with the struggle of that Self amidst adversity and violation; with the confluence of primordial images; with the unique positionings consciousness takes at the confluent streams; and with my almost instinctive urge to communicate, to speak, to write about life on the borders, life in the shadows.

Images of deep personal exploration and penetration permeate the prose and poetry sections of the text, illuminating her need to uproot the cultural constructions that are defeating to Chicana women.

Her most pronounced argument in the book is that there is a desperate need for a massive uprooting of dualistic thinking that sets people up against one another in terms of gender, religion, or culture. In the chapter entitled "Toward a New Consciousness," she writes: "the answer to the problem between the white race and the colored, between male and females, lies in healing the split that originates in the very foundation of our lives, our culture, our languages, our thoughts" (80). The New Mestiza is a multiple cultural configuration where there is acceptance for all types of individuals, a place where growth occurs and where identity is mutable and transformative. "Indigenous like corn, like corn, the mestiza is a product of crossbreeding, designed for preservation under a variety of conditions" (81). Once again, Anzaldúa uses a metaphor from nature to describe the conditions of border identity.

Anzaldúa's need for deep personal exploration and massive uprooting of destructive polemic ideologies is best exemplified in the poems "I Had To Go Down," and "Letting Go." These poems describe Anzaldúa's attempts at uprooting and reterritorializing her internal terrain through images of household dirt and grime, and the physical extraction of vermin and waste from her body. "I Had To Go Down" metaphorically chronicles the process of Anzaldúa going deep into herself to discover her own darkness and that which has been buried or neglected by describing a rare trek into the basement of her house. The basement is dark, dusty, and cold with "caked tears" on the windows that Anzaldúa scrapes off. Her fear of the basement stems from hearing footsteps and noises coming from there that sound like a "wild animal kicking at its iron cage" (167). She states, however, that she "couldn't put it off any longer. / [she] had to go down" (168). As she rummages through the cobwebs, dirt-caked pipes, and debris she nearly trips over something on the floor. It's a new root growing amidst the grime and refuse:

> A gnarled root had broken through
> into the belly of the house
> and somehow a shoot
> had spring in the darkness
> and now a young tree was growing
> nourished by a nightsun. (169)

Reminiscent of the image of the seamless sea contrasted against the rupturing land border in the poem discussed earlier, Anzaldúa uses the natural world to represent the hope and regenerative capabilities of the human psyche. Even in the harshest, most neglected spaces, new life will form and find a way to grow. The final lines reveal to us the deep psychological implications of the poem:

> Then I heard the footsteps again
> making scuffing sounds
> on the packed dirt floor.
> It was my feet making them.
> It had been my footsteps I heard. (169)

Anzaldúa's fear of venturing into the basement, therefore, was a fear of the self—a fear based on the acknowledgment of her own internalization of self-destructive border culture manifestations, such as self-hatred and confusion. She was taught to hate herself because she was brown, female, and lesbian.

The ending of this poem is quite positive despite the images of decay that characterize most of its content. Anzaldúa leaves her readers with the sense that the reshaping of her territory will now begin. Knowing that the footsteps were her own, she is free of the fear that kept her from entering the basement. The new tree growing is the beauty and strength she needed to find within herself in order to clean and remove the rubbish from her physical and psychological home.

"Letting Go" is another poem that deals with exploring the self and eradicating that which is psychologically harmful and tormenting. This poem is different from the previous one in that the metaphor is not in an external physical environment but rather one turned inside as a poison of the human body. Instead of cleaning a basement to free oneself of the harmful, in "Letting Go" Anzaldúa writes:

> You must plunge your fingers
> into your navel, with your two hands
> split open,
> spill out the lizards and horned toads
> the orchids and the sunflowers,
> turn the maze inside out. (164)

The images from nature that Anzaldúa finds within her self are both unpleasant *and* beautiful. Later in the poem, she drops "dead rats and cockroaches" alongside of "spring rain" and "young ears of corn" (164). The implication here is that complete self-exploration and purging involve all aspects of the self in order to start at a purified beginning.

The physical act of tearing one's body open to spill out the contents is curiously violent but not uncharacteristic of Anzaldúa's other depictions of self-examination. The suggestion of that violence may refer to the fact that serious self-inspection is a process that is psychologically very painful and difficult. According to "Letting Go," it is also a process that needs to be repeated. It is a "darkness you must befriend if / you want to sleep nights" (165). But once again, the poem ends on a positive level. In this case, reterritorialization has occurred, and Anzaldúa returns to a place that is comfortable:

> And soon, again, you return
> to your element and
> like a fish to the air
> you come to the open
> only between breathings.
> But already gills
> grow on your breasts. (166)

Although Anzaldúa may be playing with a rather cliché fish-out-of-water metaphor, I believe it works in the context of the poem.

Through continuous deep exploration and purging, one is able to find his/her "element" or home and better understand who he/she and where he/she belongs. Anzaldúa's psychic territorial cleansing as described in these two poems involves courage and work. But the endings of the poems, through images of nature, suggest a promise of hope and a new territory where one is at ease with their sense of self and place.

As stated earlier, inhabiting a physical and psychological territory that is not wholly one culture or another results in a sense of personal ambiguity and fragmentation that manifests itself in literature in creative ways. Anzaldúa, like Chrystos, relocates and finds refuge in nature and in a community of women that enable her to experience the sense of acceptance she needs to connect to herself and to others. Within the workings of much Chicana literature lies the theme of redefinition, which expresses the human desire for self-identification and empowerment. In her essay "Chicana Feminism: In the Tracks of 'the' Native Woman," Norma Alarcón discusses the attempt of Chicana poets to create a true self and identity by recovering and connecting with various female indigenous figures who provide a link to the past and a history for the Chicana to claim as a foundation for construction of the self in the way that Harjo does in "Deer Dancer."

As the retrieval of indigenous women in Chicana poetry appears to be prevalent, there is also a formation of the self through the creation of imaginative spiritual and supernatural female entities that represent some unresolved and undefined aspect of the author's life. By communicating with the fragmented self through these feminine spirits, Anzaldúa is able to come to terms with the undefined aspects of her life in the borderlands and to form a more complete and consciously constructed identity and terrain. Like Chrystos, she chooses who remains and who must leave her terrain and finds refuge among a community of women.

In the poem "Interface" from *Borderlands/La Frontera*, the narrator experiences the presence of an invisible, spiritual woman in her room, who gradually becomes flesh and ultimately has a love relationship with the narrator. Through the supernatural woman, Leyla, Anzaldúa confronts the dualities in the borderlands and her own sexuality that she has to redefine in order to create her own identity. In the poem, Anzaldúa skillfully sets up a parallel between the geographic and cultural border and between the spiritual and physical worlds. Although the mystical woman has apparently always been with the narrator, it is not until the narrator learns to view the world from the border, with a dual vision, that she is able to see her:

> At first it was hard to stay
> on the border between
> the physical world
> and hers.
> It was only there at the interface
> that we could see each other. (148)

Clearly, the difficulty in staying on the border not only refers to the physical and spiritual worlds but also to the border between cultures. There is always a temptation to assimilate to one culture or another in order to avoid the ambiguity, confusion, and restlessness of inhabiting an undefined sphere. But, interestingly, Anzaldúa claims that it is only on this border, "at the interface," that Leyla and the narrator can see each other. In this way, she demonstrates how it is only by disintegrating "binary dualisms and creating a third space, the in-between, border, or interstice that allows contradictions to co-exist in the production of a new element"(Yarbro-Bejarano 11). The implication is that this third realm of consciousness allows for a more sensitive vision unencumbered by cultural conventions or limitations.

Throughout "Interface," Anzaldúa also explores her sexuality. Often Chicana poets may "carve for themselves a new freedom in the treatment of sexuality. They rebel against the traditional sexual values and then depreciate the models of moral virtue" that have been socially and religiously constructed (Bornstein 43). The narrator's lesbian relationship with Leyla is depicted as loving, nurturing, and erotic. Going against the traditional stigma attached to homosexuality, Anzaldúa several times refers to Leyla as "pure sound." By equating the spirit woman with purity, Anzaldúa rejects the Catholic and Mexican cultural belief that homosexual behavior is immoral and corrupt.

The love relationship, however, is one that can only be found on the border: "We lay enclosed by margins, hems, / where only we existed" (150). Therefore, Anzaldúa believes that life on the border expands and heightens one's vision and also intensifies and frees sexual experience from social constraints. Through the creation of Leyla, Anzaldúa is able to explore these experiences and create for herself an identity in the borderlands that does not conform to any one culture but rather builds itself out of the duality and perplexity of a third, mixed-blood culture.

There are many other female figures and deities who populate Anzaldúa's text from which she gains empowerment and self-understanding. Coatalopeuh, Coatlicue, and Snake Woman all have symbolic significance connecting Anzaldúa to those aspects of her culture she chooses to embrace. She describes the Shadow-Beast as the rebel inside her that resists all forms of authority including those self-imposed, and she "is the stranger, the other. She is man's recognized nightmarish pieces, his Shadow-Beast. The sight of her sends him into a frenzy of anger and fear" (17).

Ultimately, there are *Los Chicanos* who "know how to survive. When other races have given up their tongue, we've kept ours. We know what it is to live under the hammer blow of the dominant *norteamericano* culture" (63). Anzaldúa strives and waits for the day when *Los Chicanas*, in particular, can reveal their true faces and true identities. In the light of the New Mestiza, the new consciousness, she hopes that white supremacy will be revealed for the holistic imposition that it is, that borders of all kinds, physical and psychological, will be reterritorialized and healed to resemble the seamless skin of the ocean and earth. The final poem in *Borderlands/La Frontera*, entitled "Don't Give In, *Chicanita*," addressed to young Chicanas and her younger sister, Missy Anzaldúa, in particular, leaves Anzaldúa's readers with a spirit of survival and constant resistance: "Like serpent lightning we'll move, little woman. / You'll see" (203).

Susan Griffin: Deconstructing Maldevelopment and Claiming Utopia

In *Woman and Nature: The Roaring Inside Her* (1978), Susan Griffin exposes how Western culture, through scientific, religious, and philosophic thought, has justified and defined the subjugation of women and the nonhuman natural world. Quoting famous statements and discoveries by the leading influences in Western philosophy, Griffin reveals how deeply rooted is the concept of women's inferiority. She also shows how the oppression of women corresponds to patriarchal conceptions of nature where the natural world is treated as matter created for male use and conquest. This kind of direct interrogation of foundational assumptions makes *Woman and Nature* a significant work in the field of ecofeminism.

Patrick Murphy in "Voicing Another Nature" from *Literature, Nature, and Other: Ecofeminist Critiques* discusses the double-voicing in *Woman and Nature* and claims that Griffin uses it "as one technique for exposing the illusion of objectivity and separation that philosophical and scientific discourses have attempted to perpetuate" (41). One of the voices is the voice of women in italics; the other voice is the male voice of claimed objectivity. Instead of rejecting those who have been historically responsible for the subjugation and deterritorialization of women and the landscape on her terrain as Chrystos does, Griffin invites the patriarchs along for a dialogue, to traverse the forest, the farm, and the female human body along with her.

Useful for this investigation is Vandana Shiva's definition of "maldevelopment," which, applied to Griffin's *Woman and Nature,* clarifies the salient paradoxical character of Western patriarchal development in its assumption that its ideological and industrial practices are considered "advancements" or "progress" when for women and the environment, they have been quite detrimental. It is my intention to explore Griffin's text as a primary source, going against the common practice of using *Woman and Nature* as a secondary text to support the analysis of other works, because *Woman and Nature* is a highly creative work in its own right. Like *The Way to Rainy Mountain*, Griffin's text is hard to categorize. It is historical, scientific, and poetic; it tells stories, has multiple voices in dialogue, and is deeply imaginative yet resonates sharply with truth.

Before we begin to read *Woman and Nature*, Griffin makes it clear that the work is for those who have been deterritorialized: "These words are written for those of us whose language is not heard, whose words have been stolen or erased, those robbed of language, who are called voiceless or mute" (v). Reterritorializing is a project that involves the reinscription of persons, identities, and meanings. But first, one must dismantle the notions upon which the existing identities and meanings are built. Perhaps no text more blatantly deconstructs Western patriarchal ideologies of women and nature better than Griffin's.

In "Development as a New Project of Western Patriarchy," Vandana Shiva writes about the loss of the "feminine principle" in the world order. Western patriarchal capitalistic biases against women and nature operate under the assumption that nature and women are passive and unproductive. This assumption is based on a very narrow definition of production and development, which states that "production only takes place when it is mediated by technologies for commodity production, even when such technologies destroy life" (191). Nature, therefore, is only productive when it is "developed" and commodified despite the ramifications. With production being reduced to profit and capitalistic gain, Third World women, in particular, as Shiva discusses, are thus unproductive in that their notion of productivity revolves around providing sustenance and daily provisions for survival, not procuring a financial profit. Also, women have not been constructed as technologically intelligent and capable beings, further distancing them from usefulness in a civilization of product and profit.

Shiva regards the Western concept of production as a form of "maldevelopment," which she defines as

> the violation of the integrity of a living, interconnected world, and it is simultaneously at the root of injustice, exploitation, inequality, and violence. It involves the simultaneous subjugation of nature and women. It arises from limited patriarchal thought and action that regards its self-interest as universal and imposes it on others in total disregard of the needs of other beings in nature and society. (193)

If we accept Shiva's definition of maldevelopment as true, then Griffin's *Woman and Nature* provides an in-depth history of its evolution. In *Woman and Nature*, Griffin likens nature to a woman's body, mysterious and unknown. Nature, like woman, is enticing to man and tempts him to subdue and take possession of it—to take control of land and make it useful and provide him with food. The man then claims to protect nature by using pesticides that destroy her. Women are described in accordance to the salable lumber of trees commodified and as hurricane winds he tries to control for enhancing production. He decides how the forest and the landscape should look, how the cow (woman) should be bred in order to produce the best milk, and how woman can be trained like a horse to serve man. Aspects of a horse's intrinsic character are used against her in order to subdue her—her nonaggressiveness and need for security. The body structure of a horse is not meant to carry man but strengthened to do so in order to serve him.

The commodification flows smoothly from nature to the woman's body. Menstruation is a secret to be ashamed of because it makes her unpleasant. Plastic surgery is required to make her more presentable. A clitorectomy subdues her pleasure so she is not distracted from her work. Then the womb is mined for what man can prostitute.

Griffin uses these examples to demonstrate how man has separated himself from woman and nature and separated the woman from herself: "Her will from her body. The knower from the known. The speaker from the mute. Self from self. From the nocturnal. From the nightmare. Discovery from the dream. Her will from her body" (97). Although she portrays such a closeness between women and nature in terms of their shared experiences under patriarchy, Griffin does not propose an essentialist relationship between the two. Karla Armbruster notes, "the way this text juxtaposes scenarios of the oppression of women and of nonhuman nature throughout the history of Western civilization highlights the connections between the *cultural* positions of women and natural entities" (21). This positionality is at the heart of the study of reterritorialization and its quest to reposition and relocate those who had their positions chosen for them.

Griffin also shows how man further constructs separations and boundaries—silver from lead, pelt from fox, spirit from body. Griffin's portrayal of courtship is that the man wins over the woman by professing love, then controls her by setting up boundaries, a system of measurements—miles, feet, gallons, hours, epochs, minutes. His certainty about what he does is proved through numbers; everything is mathematically calculated and exact.

Through this cultural maldevelopment that exploits and reduces women and nature to dehumanized objects for the patriarchal enterprise, women thus learn to hate themselves and each other. At the end of the book, Griffin describes man in terror at being able to identify with woman while woman comes to the realization of how she has been defined:

> And then another thought came upon him, so terrible he could scarcely hold on to it. Suppose there is no difference between them except the power he wields over her. And suppose that in an instant of feeling himself like her, he let this power go, then would he not become her, in his own body even. And some part of him seemed to know what it would be to be her in his body, and how he came to know this he does not choose to remember. (150)

In an epiphanic moment, man has come to realize that the boundaries he has constructed between man and woman and man and nature are disintegrating, and the sudden vulnerability he faces is horrifying. After this realization, Griffin describes man rejecting these thoughts and reestablishing his space of superiority "as far as it could go to the outer limits of the universe" (150). In this space he begins uttering measurements to woman of how far galaxies are from the earth and how fast they are moving away as if to increase his own psychic distance from his terrifying realization. But this chapter, appropriately named "Terror," ends with man still living in fear at the recognition of the notion of mystery, at the final acceptance that much of the universe is "still unknown" and, therefore, in a position of power over him.

Woman then becomes aware of her strength and power. She gets back in touch with what is still wild within her and recognizes that she is a part of nature and that is her power. Also, by recognizing the earth as her place of existence, woman embarks on a journey of rebirth and rediscovery: "This place. This place in which she breathes and which she takes into herself and which is now in her, sleeping inside her. What sleeps inside her? Like a seed in the earth, in the soil which becomes rich with every death," the strength inside her builds and the healing within her own body and mind begins (167). That man has separated himself from nature is his doom, and the connection continues to terrify him.

The utopian ending of *Woman and Nature* involves entering a new space as described in the chapter, "The Opening." This new space is described by the italicized female voice as one "which is never separate from matter" and "the shape of experience" (169). Interspersed with these descriptions are those by the assumed voice of objectivity, which continues to try to impose on woman what her space consists of and what characterizes it. When woman asserts that it is a "space in which there is no center," the voice responds with, "space filled with her disintegration." When woman claims that it is a space "where all certainties change," the voice responds with, "space in which she feels she is coming apart" (170). For the female speaker, the change and the mystery are positives in her space and the source of her power. For the voice of objectivity, which still retains its reliance on absolutes, the new space she is creating is filled with disorder and chaos that could only lead to her ruin.

The ending of the work presents women who feel for and identify with earth, loving nature with no desire to subdue it. They recognize they come from the same soil and are empowered with this knowledge and community they share. The chapter "Transformations" most directly expresses the metamorphosis that has taken place among women:

> We are no longer pleading for the right to speak: we have spoken; space has changed; we are living in a matrix of our own sounds; our words resonate, by our echoes we chart a new geography; we recognize this new landscape as our birthplace, where we invented names for ourselves; here language does not contradict what we know; by what we hear, we are moved again and again to speech. (195)

This new landscape and new geography created by the women form a place of their own making where they are renewed, reinvented, and empowered to speak for themselves. As an activist text, *Woman and Nature* demands the relationship between men and women, and that between humanity and nonhuman nature, be reconsidered and transformed. The subjugation of nature has resulted in the prevailing ecological crisis, and the subjugation of women has resulted in a continuing debasement of women, in which domestic violence, rape, anorexia nervosa, lower salaries, and unrealistic standards of beauty are serious symptoms. One of the goals of ecofeminism as expressed by Ynestra King in "Healing the Wounds: Feminism, Ecology, and the Nature/Culture Dualism," is to genuinely and actively seek an antidualistic philosophy in Western culture as another stage of human evolution called "rational enchantment." This "rational enchantment" will involve "a new way of being human on this planet with a sense of the sacred, informed by all ways of knowing—intuitive *and* scientific, mystical *and* rational" (120). New definitions of "progress" and "production" need, therefore, to be conceived along with new ways of perceiving women and nature. Susan Griffin's *Woman and Nature* successfully exposes those destructive assumptions and cultural meanings that have caused violence toward women and the environment so that the healing can begin and continue.

The project of Western patriarchy thus far has been to construct a civilization in which it has been understood that antithetical epistemologies cannot coexist. This paradigm is rationalized by those in power in order to maintain their position and to sustain the subjugation and oppression of women, nature, and marginalized cultures. One cannot stifle, however, the human desire for freedom to belong—to oneself, to others of one's choosing, to a certain place. One also cannot stifle the human need and quest for safety, to live on a sustainable planet without worrying about what is in the eight glasses of water we should be drinking every day.

Place is not only physical, but also ideological. Places have meanings as well as geographical locations. Chrystos's text is a battlefield and a place for her to declare war on injustice. The borderlands for Anzaldúa are concentrated with multiple meanings, and for Griffin ideological constructs of woman and nature are unearthed; the woman's body and natural terrain become places to dismantle patriarchy. Fundamentally, all three of these writers are declaring war on hostile ideological contact zones. They demonstrate how polemical the dialogue has become in the areas of race, gender, and nature and are not afraid to confront and reject destructive ideologies and their followers, inside and outside their texts. As they confront the injustices that have alienated them from themselves and the surrounding world, they are remapping a psychic terrain in order to survive in a world of their own making.

Chapter Four

Environmental Reterritorializations: Reinhabitory Writings

> The earth is no wanton to give up all her best to every comer, but keeps a sweet, separate intimacy for each.
>
> —Mary Austin, *The Land of Little Rain*

> We once more know that we live in a system that is enclosed in a certain way, that has its own kinds of limits, and that we are interdependent with it.
>
> —Gary Snyder, "Reinhabitation"

In an essay from *A Place in Space* (1995) entitled "Reinhabitation," Gary Snyder defines what he considers a "reinhabitory" relationship with the nonhuman natural world. He claims that "reinhabitory refers to the tiny number of persons who come out of the industrial societies (having collected or squandered the fruits of eight thousand years of civilization) and then start to turn back to the land, back to place" (190). He maintains that reinhabitory living is a way of life that entails intimate association with a place and a recognition of its bioregion and the interdependence of all living things within that place. Choosing to live an inhabitory life is not only a physical decision but a moral and spiritual choice as well. Snyder describes the expression of the spirituality as "feeling gratitude to it all; taking responsibility for your own acts; keeping contact with the sources of the energy that flow into your own life (namely dirt, water, flesh)" (188). The expression of this ethical and spiritual relation with the land also involves an awareness of the land's history and a respect for the wisdom it contains. Writers Wendell Berry and Linda Hogan are reinhabitory writers in that they articulate in their writings the kind of physical, moral, and spiritual existence with nature that comprises Snyder's definition of inhabitation. Berry and Hogan pay a great deal of attention to the workings of the natural world, and its processes are part of their daily life and existence.

The next question at hand is how is it that these writers are reterritorializing their environments in their writings? Like the other writers we've reviewed, Berry and Hogan are redefining meanings in relation to place. This chapter, however, deals more directly with how the writers redefine their environments. In a work of creative nonfiction called *Dwellings: A Spiritual History of the Living World* (1995), Hogan reinterprets the way Western culture has codified various creatures and places from the natural world with negativity. Berry poetically reconfigures his relationship with the farm landscape into a historical interaction and close, reciprocal connection like that of a marriage in *Clearing* (1974). His volume of poetry entitled *The Country of Marriage* explores this kind of connection as well. Both writers reveal their environments as extensions of the self, demonstrating the interconnections between the human and natural worlds and how human processes throughout life and death mirror those of the land. Their interactions with their environments provide opportunities to learn more about themselves and the world around them while continuing to nurture an abiding spiritual connection with nature.

Linda Hogan: The Terrestrial Intelligence

Linda Hogan, as mentioned earlier in chapter two, is an American Indian writer of Chickasaw descent who has published novels, poetry and nonfiction prose. In the preface to *Dwellings*, Hogan reveals that the writings have grown out of her "wondering what makes us human, out of a lifelong love for the living world and all its inhabitants" (11). She also states that the work reflects "the different histories of ways of thinking and being in the world" and that she writes "out of respect for the natural world, recognizing that humankind is not separate from nature" (12). Her devotion to place reflects these inspirations and requests our acknowledgment of the planet we call home and its nonhuman communities. The writings in *Dwellings* have also grown out of Hogan's "native understanding that there is a terrestrial intelligence that lies beyond our human knowing and grasping" (11). This "terrestrial intelligence," although acknowledged by Hogan as inaccessible to human comprehension, plays a significant role in inspiring *Dwellings* as a spiritual history where oral traditions and nature's mysteries are given prominence over Western ideological constructs of nature that have been confirmed as detrimental in the midst of the contemporary environmental crisis. Informed by her Native heritage that encourages reverence for, and reciprocity to the natural world, Hogan's respect for the earth's terrestrial intelligence is clear in her insistence for a more balanced relationship between the spirit world and human world. This balance is essential in maintaining a sustainable planet and the answer to much human suffering.

Hogan's terrestrial intelligence is not confined to territories of solid ground, as we gathered from our earlier discussion of her work. In *The Book of Medicines* (1993), water becomes a recurring image for physical and psychological healing in a contemporary world of sexism, drought, violence, and hunger. In this poetic work, Hogan evokes terrains reminiscent of tribal origins, such as the Chickasaw emergence story of rising from an underground origin into the world through a lake. Additionally, Hogan attempts to integrate the past with the present in this work and to reconcile her mythic/historical sense of place with her contemporary one. These two conflicting territories give rise to an aquatic, poetic terrain, which is at once beautifully nostalgic and, at the same time, sorrowful and brutal. Like *Dwellings*, *The Book of Medicines* challenges Western constructs but uses an ecofeminist activism that brings together women and water imagery to expose male exploitation of women and nature on an aquatic terrain. Hogan's belief in a terrestrial intelligence or faith in the spirit and mystery of nature as a source of wisdom is evoked in this work as an aquatic intelligence. In both works, social activism is drawn from sources of spirit and mystery of nature where injustices committed against women and the nonhuman natural world are addressed. Hogan demonstrates that by studying this form of intelligence, we can learn how to take better care of our environment and one another.

We should note her different perspectives on being in the world and her philosophies in relation to nature. Her focus on the spiritual dimension of the dwelling places of living creatures reinforces the notion of the earth as a vital, living organism upon which we live and also connects the earth to a larger cosmic realm.

The small, seemingly trivial dwellings of creatures are as spiritually significant for their existence and function in the natural process as the whole of the earth itself. Her questioning of Western meanings imposed on aspects of the natural world force us to recognize the anthropocentric constructions of nature as an ideology with many fixed meanings. But *Dwellings* is more than a quest redefining meanings and places, it is also a guidebook on how to nurture a more spiritual connection with the world and its inhabitants.

Linda Hogan's approach to nature is nothing less than mythical, mystical, and magical. Much of her philosophy and view of nature is founded in American Indian mythology. She claims that "in recent times, the term 'myth' has come to signify falsehood, but when we examine myths, we find that they are a high form of truth. They are the deepest, innermost cultural stories of our human journeys toward spiritual and psychological growth" (51). The symbolic meanings of animals and places within the natural world that she uses to dismantle existing Western notions come from these myths that shape her worldview.

In the chapter "The Bats," Hogan describes these winged creatures, who are often associated with the blood-sucking terrors of Dracula, as sacred creatures that occupy two worlds giving them great insight and wisdom. These bats "live inside the passageways between earth and sunlight." They are "two animals merged into one, a milk-producing rodent that bears live young, and a flying bird. They are creatures of dusk which is the time between times, people of the threshold" (27). For Hogan, bats have spiritual significance in that they exist in a liminal state between worlds and therefore act as guardians of the passage into a higher spiritual state. Hogan claims that in Native stories "the bat people are said to live in the first circle of holiness. Thus, they are intermediaries between our world and the next" (27). As intermediaries, bats transcend the Western and popular culture stigmas that have associated them with fear and evil. Also, Hogan's conception of the next world of spirits is not one of evil and death but a world populated by holy beings and ancestors. The bats, therefore, are guides to a more spiritual existence, not creatures from a horrifying darkness.

Hogan's chapter on snakes is similar. She deconstructs the association of snakes with Satan as the force that tempts Eve out of her virtue and redefines the serpent as a symbol of wholeness and regenerative life. Due to the snake's ability to coil itself in the form of spiraling circles, it echoes the circular life philosophy of continuity, reciprocation, and holistic living (nurturing spiritual, mental, physical, and emotional needs) rather than the Western linear construct, which leaves a loose end dangling into oblivion.

In the chapter "Creations," Hogan writes that "unlike the cyclic nature of time for the Maya, the Western tradition of beliefs within a straight line of history leads to an apocalyptic end. And stories of the end, like those of the beginning, tell something about the people who created them" (93). Hogan is suggesting here that people of Western cultural backgrounds live in destructive ways that will ultimately lead to their own doom. By living life in a linear fashion, one takes without giving back and progresses toward a goal without examining future consequences. Linearity suggests that there is a beginning, middle, and an end, and one lives in accordance with that and moves toward that end. A linear existence may involve engaging in various forms of self-destructive behavior as well as behaviors that are destructive to others or to the nonhuman world we inhabit. Hogan believes that when people take responsibility for one another as well as the earth and perceive life as circular with transformations instead of conclusions, their attitude toward life prioritizes preservation of what they may someday need.

Hogan contends that "Before SNAKE became the dark god of our underworld, burdened with human sin, it carried a different weight in our human bones; it was a being of holy inner earth" (140). As a being associated with the inner earth, the snake is at the core of life and creation for Hogan and many other Native cultures. In Gloria Anzaldúa's *Borderlands/La Frontera*, the serpent's mouth is symbolic of womanhood: it is considered "the most sacred place on earth, a place of refuge, the creative womb from which all things were born and to which all things returned" (34). It is intriguing that the snake is a symbol of life and birth in Native oral traditions, but in the Western tradition, it is the symbol of sin and death. Hogan maintains that "in more recent times, the snake has symbolized our wrongs, our eating from the tree of knowledge, our search and desire for the dangerous revelations of life's mystery" (141). The biblical story of Adam and Eve has indeed given snakes a bad reputation. The edenic snake is the ultimate symbol of temptation and represents the fall of mankind. Hogan's work provides a much-needed alternative symbolic meaning for this creature.

In her attempt to reshape our notions of bats and snakes, Hogan strives to break down the human/nature dichotomy and heal the alienation between humans and the natural world that has led to environmental degradation. She maintains that at the core of this alienation are Western religion and philosophies. In "Creations," Hogan asserts that "the Western belief that God lives apart from earth has taken us toward collective destruction" (85–86). In many Native philosophies, gods and goddesses, who represent Hogan's idea of a terrestrial intelligence, walk amongst the people in spirit form or in the shapes of animals and other people. The fact that the Christian God is located in a separate "heavenly" realm far higher than the earth and its inhabitants suggests that the earth is a place of sin and physical matter that needs to be transcended. The body dies and remains on earth to disintegrate while the soul transcends to a spiritual, unearthly place. While followers of Emersonian thought and earth-based religions have worked to show the presence of spirituality in nonhuman living things, the separation of God from earth suggests that inhabitation of earth is ultimately undesirable.

In *Dwellings*, Hogan explores in detail the homes and dwelling places of animals, birds, and Native peoples. By exploring the history of the natural world in regard to its Native mythological and spiritual significance, she highlights how close the lives of animals are to human life in order to combat the human/nature alienation created by Western thought. In one instance, she tells a story about how one day, hiking up a mountainside to hear the voices of great horned owls, she discovered in front of her a nest that had fallen from a tree along her path. Much to her delight, she discovered that the birds that built the nest used a thread from one of her skirts to build their home. She states, "I liked it, that a thread of my life was in an abandoned nest, one that held eggs and new life" (124). On looking at it more closely, she discovered strands of her daughter's hair were also used to build the nest. Hogan claims, "I didn't know what kind of nest it was, or who had lived there. It didn't matter. I thought of the remnants of our lives carried up the hill that way and turned into shelter" (124). What was most important for her was the fact that the birds benefited from her existence in some way, and that she played a small role in the issuance of new life. In this case, Hogan uses the example of the bird's nest to point out that many creatures make use of the materials they find in the wild to structure and build their dwellings, but profligate use of materials seems to be a characteristic more analogous to human consumption.

In addition to birds' nests, Hogan depicts caves as important dwelling places for Native communities. First, caves provided the entranceways between worlds; these are the homes of bats, who act as intermediaries. Second, caves were refuges from dispute and strife that took place in the world outside the caves: "In earlier days, before the springs and caves were privately owned, they were places of healing for Indian people, places where conflict between tribes and people was left behind, neutral ground, a sanctuary outside the reign of human difference, law, and trouble" (29–30). The caves, therefore, provided a temporary retreat for the tribes from war and conflict. They were places to go to think and regroup, to rest and plan the next course of action. In Hogan's novel, *Mean Spirit* (1990), the character Michael Horse retreats to a cave to record the history of his Native people in a journal in an effort to supplement the Bible with Indian ways. Character Belle Graycloud also believes that this "Sorrow Cave" is the home for the bats that bear powerful medicine to those who believe. For Hogan, caves are places of great spiritual significance: safety and sanctuary, not dark holes where fire-breathing dragons await curious travelers as in the Western literary tradition.

She also assigns a distinctly female symbolic meaning to caves. Once, on a journey near the Continental Divide with her family as a young girl, she saw an African lion at the mouth of a cave. She told her father, but he did not believe her because he did not see the lion. Although Hogan knew that the lion was there, her father was unable to locate it even though he returned from the cave smelling of the lion. It became clear to Hogan at a young age that there are places on earth where men do not belong. These are places where they either do not have access or do not have knowledge of where they are and what surrounds them. Although Hogan admits that a terrestrial intelligence lies beyond human comprehension, this story suggests that perhaps there are places where women may catch a glimpse of the intelligence while men remain unaware of its presence. Hogan states that "caves are not the places for men. They are a feminine world, a womb of earth, a germinal place of brooding. In many creation stories, caves are the places that bring forth life" (31). Many creation and emergence stories tell of the tribe coming into being from the underworld through some passage such as a cave. The underground tunnels and caves, therefore, could be equated with the womb and birth canal, where the people emerge from the world of darkness into the world of light.

One of Hogan's fundamental calls in *Dwellings* is for readers to teach and demonstrate stewardship for the earth. It is a practice necessary for preserving the earth's creatures and healing the gap between the human and nonhuman world. She writes that "caretaking is the utmost spiritual and physical responsibility of our time, and perhaps that stewardship is finally our place in the web of life, our work, the solution to the mystery of what we are" (115). Until we learn "our place at the bountiful table, how to be a guest here, this land will not support us, will not be hospitable, will turn on us" (46). This prophetic declaration debunks the anthropocentric notion that the earth is here to serve human needs, and places the human world on equal footing with the other "guests" at the table, the nonhuman guests. Also, emphasized in this passage is the spiritual connection with the earth and its terrestrial intelligence. Hogan passionately and convincingly suggests an explanation for the mystery and meaning behind human existence—recognition of one's place within the ecosystem and responsibility for its well-being. She practices what she preaches. Near the end of *Dwellings*, we discover that she works in a rehabilitation facility for birds of prey. Clearly, she believes that stewardship is our responsibility as co-inhabitants of this place we call home.

In her most direct attempt to contest the manner in which human beings view nature, she writes:

> We are of the animal world. We are a part of the cycles of growth and decay. Even having tried so hard to see ourselves apart, and so often without a love for even our own biology, we are in relationship with the

> rest of the planet, and that connectedness tells us we must reconsider the way we see ourselves and the rest of nature. (114–115)

Reading Hogan's *Dwellings* can make one reconsider this relationship. Her book, with all its stories, re-creates the life of the natural world that has been objectified, and it redefines nonhuman creatures that have been negatively stereotyped. She provides an alternative way to conceive of snakes, bats, birds, caves, nature, and the human world in relation to the nonhuman world.

Additionally, Hogan discusses topics and issues surrounding ecopsychology, animal rights activism, hatred of wolves, water pollution, the Mayan civilization, and other ancient peoples who were inhabitants of the land to illustrate her ideas. In doing so, Hogan challenges prevailing destructive views toward the natural environment by decentering human beings and placing them at the guest table alongside all other life on earth.

Wendell Berry: Rewriting the Farm's Narrative

Wendell Berry's home is a farm on the Kentucky River. Like Hogan, he explores environmental issues in novels, essays, and poetry, and his actions support his ecological philosophies. "His work is of place, and whether he writes poems, novels, or essays, he reveals his preoccupation with the land and his sense of culture as derived from his acceptance of a way of life many people today regard as 'alternative,'" claims Leon Driskell (*Dictionary of Literary Biography* 62). Berry lives an alternative lifestyle by living closely with the land and exploring the relationships between the human and nonhuman worlds in his writing. In exploring these relationships, Berry rewrites the farm landscape into a narrative of marriage, history, and community while he provides a lesson in stewardship in order to save the farm from ruin. Ultimately, the construction of the farm and his work on it become a construction of the self as Berry prepares himself for pending life changes.

In his volume of poetry *Clearing*, Berry employs a poetic language that is concrete yet mystical as he communicates that self and community involve more than that which is present and visible at a given time. Berry is acutely conscious of the land's past and how it connects to himself and other creatures that inhabit his farm. The book's structure consists of seven long linked poems that progress from describing the history of the farm ("History" and "Where") to describing his work of caring for the land and clearing the farm in "The Clearing" and "Work Song." The volume then concludes with "The Bed," "From the Crest," and "Reverdure." These poems indicate a turning point in Berry's life when the presence of death launches a lyrical self-preparation for becoming part of the cycle of life.

In the first poem, "History," Berry writes of his initial meeting with the farm landscape and how it had been left to decay by previous owners. The farm, he claims, had been passed from one owner to the next until a developer bought it and instigated its ruin. This developer, described by Berry as having no respect or reverence for the land aside from its monetary possibilities, is linked to a city environment. Berry describes his mind as a polluted "disordered city" that survives by overconsuming valuable resources (18). Although this metaphor may exercise a rather idealistic dichotomy separating the country from the city, the point that Berry makes involves the excess of industrial consumption that doesn't see beyond its immediate need for material gratification.

In an article entitled "Forever Virgin: The American View of America," Noel Perrin writes about this blind consumption using the metaphor from *The Great Gatsby* of the green light that blinks from across the bay. As the narrator turns his back on the green light, he visualizes what it must have been like for the first settlers who came to America and saw the green of the northeast coast. The settlers saw the landscape as a fresh, untouched, virgin world, where the future is open and green like the breast of mother nature. If we include animals as inhabitants of the land, then the land is not untouched and somebody always lives where people settle; therefore, someone is always being displaced from their home. The problem, Perrin states, is that Americans still view this country as an inexhaustible and ever-nourishing maternal provider; we still see the green light, signaling to *go ahead* and consume. Remnants of this buy-and-consume practice of the landscape is apparently what Berry found to be the most recent happenings in the history of the farm.

But Berry not only wants to reclaim the physical landscape of the farm to recover it from its battered condition, he also wants to redefine notions of history that leave out the daily lives of those who have chosen to inhabit the land. Much of what we tend to think of as historical has taken place in government buildings and state capitals, where, Berry states, "the accusers have mostly been guilty" (12). There is as much history taking place where the constructors of history have not taken enough time and care to look. Referring to the courthouse documents concerning the farm, Berry knows that the land contains its own history that is often not recorded (12). "Nature writing" or place-based literature, which includes novels, poetry, essays, personal accounts, and diaries, reveals the mind of the people who inhabit a landscape and provide a history and specificity of place that is unavailable in court documents or environmental history textbooks. Works such as *Clearing* or Hogan's *Dwellings* supply the personal element that expresses the minds of the inhabitants and their interior interactions with their place.

As mentioned earlier, Berry is very aware of his own sense of habitation on the farm, and he envisions those who came before him as part of the farmscape and, therefore, part of himself:

> All the lives this place
> has had, I have. I eat
> my history day by day.
> Bird, butterfly, and flower
> pass through the seasons of
> my flesh. (5)

Through his attachment to place, Berry forms a community with the dead inhabitants who have passed through the cycle of life before him to become part of nourishing the soil. It is clear from this passage that Berry's view of nature and place is nonanthropocentric and celebrates the interdependence and interconnectedness of all living things in the natural and human worlds.

Defying notions of capitalistic ownership, Berry recognizes that the farm is not just his land to nurture into production but rather a land that belongs to a larger, mysterious realm—a land upon which other humans, plants, and animals have lived, worked, traveled, and died. His connection to this land is his connection to the other living creatures that tell the story of the farm.

In *Clearing*, Berry writes a new narrative for the farm, one that does not leave behind thoughtless owners or nonhuman inhabitants of the past but becomes part of their foot trails and accepts both their positive and negative aspects, which have informed the farm and its history of "greed and innocence, / its violence, its peace" (6). Still, the story and new narrative, which Berry put into poetry are of a new farmscape, one that has been cleared of overgrown bushes, trash dumps, and old fence rows. Berry is the caretaker who regenerates what was dead into the cycle of life and rebirth. A dead elm tree, which Berry burns, turns into white ash and is carried in the wind to spread about the land, nourishing the surrounding plant life (29–30).

Berry writes of his relationship to the land as comparable to a marriage, requiring care and fidelity as the old and new come together and the lives of the present and past join to mate (47). The dichotomies of man and nature, life and death, light and dark, heaven and earth all seem to be enjoined in his poetry. Reviewer David Ignatow comments that Berry's devotion to the earth is so profound "in that one gets the feel in his poems of a lover and his beloved in mystical union" (317). The marriage is not just between himself and the place but also between himself and his work on the farm. The act of clearing and nurturing the land is a spiritual and therapeutic process for Berry. Admirably, it is a love for place and work that is unconditional. The beautiful farm that is "half-ruined" and the work that is "wearing" are accepted with all their imperfections and difficulties. It is very important to mention, however, that "this marriage, blessed and difficult" also refers to the marriage to his wife who lives with him on the farm. Although she isn't mentioned directly in the poems, she is part of the "we" community on the farm that he mentions frequently throughout *Clearing*. He also addresses her in his essays and occasionally in other volumes of poetry.

If the relationships between human and nature and nature and nature are enjoined like a marriage in *Clearing*, then perhaps all of the components of the farm, including Berry, and the land work together in one system where they do not oppose one another but rather enable one another to exist.

> The farm must be made a form,
> endlessly bringing together
> heaven and earth, light
> and rain building, dissolving,
> building back again
> the shapes and actions of the ground. (42–43)

And this system proceeds in a continuous cycle that involves life, work, death, and resurrection. Also interesting here is Berry's notion of the farm being made "a form." Berry doesn't view the landscape from a perspective that singles out its wildness or its tumultuousness and unpredictability. Rather, there is a form, a system, a cycle, a regularity that oversees all that seems unruly in the world of nature—an ecosystem. There is also a sense of a larger cosmic and spiritual realm that maintains the order evoked throughout the work by Berry's mysticism. Berry's faith lies in his knowledge that after death the soul returns to the earth, "returns to the wild / where nothing is done by hand" but rather by a higher, more divine power which oversees the chaos and sustains life through form (43).

Berry's intention to "form" and reterritorialize the farm stems from his desire to restore the farm's ecological health and to establish his own sense of self in accordance with place.

For Berry, the farm is not only a means of providing subsistence but also a form of self-expression. As an art, the farm gives Berry that which he needs to exist creatively in the world by forming and shaping the farm in accordance with his own vision. In fact, vision is the central theme in the title poem "Clearing." In this poem, vision is the chainsaw's edge that severs off the neglect from the landscape, and it is the vision that sees the farm healthy and thriving with wildflowers and woodland.

The poem "Clearing" with its descriptions of removing the wasteful and useless to replant new life is no doubt a metaphor for Berry of self-reflection and personal change. The abandoned farm represents those aspects within himself that he has neglected or which have brought him harm. Like the other works we have examined, the act of reterritorializing the external environment becomes symbolic of the human desire to change.

There are several aspects Berry is trying to change in himself throughout the writing of *Clearing*. In the poem "From the Crest,"

Berry states that his "life's wave is at its crest" (42) and later, that "the grave is in sight, / the soul's last deep track / in the unknown" (43). Berry, in all his cherishing of the land and its cycles and communities of past inhabitants, is recognizing that he is approaching his own death and will soon cross into the next phase of the cycle of his life. He is preparing himself to become a part of the community of the dead whose lives still impact the soil and mark the land. While he bears respect for the cycles of nature throughout the volume, present in this passage is a man trying to get used to loss. He is preparing himself for the journey where he will give up that which he has known to return to the wild. In this preparation he strives for acceptance of this change and patience for its arrival.

It is at this point that the reader of *Clearing* begins to understand Berry's preoccupation with the life/death cycle and his view of the landscape as a history and expression of its past inhabitants. The focus is no longer on the reconstruction of the farm, but on the reformation of the self in preparation for death. In writing the farm's new narrative, Berry writes a new narrative for himself. He spends the course of the book reconnecting himself to nature in order to make the progression toward old age and death an easier transition. While it seems true that Berry mourns the thought of passing from human existence, his love for the land and its natural processes are a comfort to him. He physically and spiritually accepts the interconnectedness of all things and recognizes he is a guest at the table and part of the ongoing cycles. As with any circular process, there is always a return to an altered form of the present state. Berry knows that his life in one way or another will experience a rebirth in the natural world that he cherishes.

Vernon Young claims that Berry is also struggling to produce unity in a life that has been divided by conflicting devotions. He contends that Berry is "anxious not only for the future, which will not be in his hands; [but also] anxious for the present in which he is being devoured by his own commitment" to the farm (35–36). Given Berry's extraordinary devotion to the farm and his consuming passion for its history and past inhabitants, Young's theory is certainly worth exploring.

In the poem "Work Song" Berry addresses this precise fear in section three entitled "Passion." In this section of the poem, he describes his intense love for the land, which has played a significant role in determining his path in life. Berry describes coming to the farm in terms of a lifelong passion, although at times unconscious, which had led him there. In the meantime, he had become a professor and a successful writer. After becoming preoccupied with the farm and his work of clearing it of refuse, Berry wonders about his ability to write and whether his passion for the farm has consumed that ability:

> Can it lead me away
> from books? Is it leading me
> away? What will I say
> to my fellow poets
> whose poems I do not read
> while this passion keeps me
> in the open? What is
> this silence coming over me? (33)

For someone who has built a successful career out of a love for books and writing, silence can be a harrowing experience. Although it is a very unsettling fear for Berry, it seems that his passion for the land does not lead him away from his writing vocation but feeds and informs it. His love for the farm and passion for the natural world is poignantly obvious in *Clearing*, as it is in his other works. Berry is able to resolve these dividing passions by writing about the land and using his experiences on the farm to teach others stewardship and appreciation for the earth's gifts.

He is also able to resolve his commitment to each vocation in a very natural, cyclical way. He writes in the winter and farms in the warm seasons. In the final section of the last poem in the volume, "Reverdure," we discover that he writes and farms in accordance with the seasons. The words and the writing sustained him through the winter months, and now that his home is surrounded by the sights and sounds of spring, Berry puts his books away and returns to working the land. His musings of the land's history and its now-dead inhabitants are ended along with his fears of the future and dying.

Berry lives fully in the present when he is working on the farm. The winter months are reserved for self-reflection and examining his relationship with the world around him. In writing *Clearing* during the winter, he entertained and came to terms with certain fears while he perhaps planned and organized what needed to be done with the farm come spring. And he ends the volume with the beginning of that sacred work.

In *Clearing*, Berry rewrites the story of the farm and himself. He acknowledges and learns to accept that his life is at a turning point, and he reterritorializes his space and mind in accordance. The new narrative also rewrites the old story of when the farm was owned by proprietors who abandoned the land and left it in ruin. But the narrative is also of larger importance. It is a new narrative for the way human beings should relate with the natural world around them. By showing the close interaction, mutual respect, and care for the land in terms of one's own marriage, ancestors, or family community, Berry redefines the way the human world perceives nonhuman life.

Both Linda Hogan and Wendell Berry exhibit, in their works and in their daily lives, the ethical and spiritual approaches to the nonhuman world of a reinhabitory existence. Gus diZerega supports this ethical approach to the natural world in his article "Individuality, Human and Natural Communities, and the Foundations of Ethics." In his discussion, diZerega compares several kinds of communities—the political, the family, and the ecological community. He states that "creating strong vibrant human communities in no way removes us from the ecological community, and its prudential and ethical implications with regard to our actions" (36). DiZerega concludes by urging people to perceive and accept the ecological community and treat it with the respect and responsibility they would give to those in their families or communities, which is precisely what Berry and Hogan do in *Clearing* and *Dwellings*. There are laws protecting human rights, and there should be more laws protecting the rights of the nonhuman communities.

DiZerega claims that "the more all encompassing and intimate the character of our community relationships, the stronger the individual obligations" to help protect and preserve those communities (34). Berry and Hogan have intimate connections with the nonhuman natural communities. Both are caretakers demonstrating the responsibility and stewardship we all should confirm in our own actions in relation to the land and its animals. They also both describe the relationship between the human and nonhuman world as a partnership that should be built upon respect and reciprocal exchanges. What perhaps makes them different is that Hogan's perception of nature has a mythical basis; it is informed by the history and spiritual dimensions of the oral tradition that intimately connects elements of nature to specific gods, goddesses, symbolic meanings, or cosmic events. Berry's spirituality regarding the landscape is characterized by a less determined, cosmic order but one that still undeniably exists. He is also more apt to impose his will upon a landscape than perhaps Hogan would be. His act of clearing the farm in order to re-create his vision may be to save the farm from ecological disaster, but it is an act of domination nonetheless.

Before Hogan and Berry are able to change the way humans act and behave in relationship to the natural environment, they have to change the way humans think about nature. Once this change occurs, perhaps stewardship will follow. Berry and Hogan demonstrate, through their writings of place and meanings, that there are ways of thinking about and interacting with nature that are nurturing and preserving rather than abusive and destructive. For both Hogan and Berry, this way of thinking emphasizes drastic unlearning of Western cultural alienation from nature, and encourages acknowledgment of and respect for the soil from which we came, to which all will ultimately return.

Chapter Five

Reterritorialization and American Indian Activism

> The differences between tribal imagination and social scientific invention are determined in world views: imagination is a state of being, a measure of personal courage; the invention of cultures is a material achievement through objective methodologies. To imagine the world is to be in the world; to invent the world with academic predications is to separate human experiences from the world, a secular transcendence and denial of chance and mortalities.
>
> —Gerald Vizenor, *The People Named the Chippewa*

For thousands of years, American Indians held the lands of the Americas in trust. How has it come to pass that the peoples who traditionally sustained themselves directly from the earth have been witness to the most severe levels of industrial waste and environmental exploitation polluting the second half of the twentieth century? Water tainted with industrial waste, strip mining, radiation exposure causing illness or even death, toxic goundwater killing livestock or crops, and Native lands targeted for waste dumps and landfills are just a few of the problems Native peoples face in their homelands. The writers I have discussed thus far are to some extent social and/or environmental activists, but environmental destruction has been alluded to only on the periphery or as a metaphor or relational cause to help support other avenues of social protest or personal exploration. Simon Ortiz, Wendy Rose, and Gerald Vizenor openly address environmental injustices as a form of action to create awareness of historical and present-day injustices as well as to present future apocalyptic possibilities.

Simon Ortiz, in *Fight Back: For the Sake of the People, For the Sake of the Land* (1980), shares the history of the Acoma people, who experienced the boom of the uranium mine industry. He speaks out against the exploitation by anti-human capitalists, shares the experiences of those who suffered physically and psychologically while working in the mines, and raises a voice of protest against continuing forms of environmental racism. Wendy Rose in *Bone Dance: New and Selected Poems, 1965–1993* (1994) raises a protesting poetic voice against archaeological digs that unearth and desecrate the bones of her ancestors. And Gerald Vizenor in *Bearheart: The Heirship Chronicles* (1978) presents a post-apocalyptic environmental wasteland that seems horrifyingly possible. For the advancement of the present environmental movement and the development of ecocriticism, these writers are significant in their ability to expose the subtle (and not so subtle) forms of human suffering and displacement that continue to engender alienation and abuse of the nonhuman natural world which circle back and ripple out, perpetuating estranged relations among human beings.

The purpose of this chapter is to illustrate other forms of activism and reterritorialization in relation to American Indian authors and to underscore that the three forms applied previously in this journey—mythic, psychic, and environmental—are not prescriptive forms of relocation but areas of emphasis. The boundaries between each were blurred in the earlier chapters and will be equally imprecise in this chapter. Technology is the target of Ortiz's activism. Wendy Rose is an academic activist who attempts to redefine the colonial attitudes of anthropologists and archaeologists, while Vizenor's activism is ideological in that a horrific fate meets those characters whose ideologies remain fixed and static. At the core of these works is a protestation against dehumanization and exploitation.

Simon Ortiz: Uranium Mines and the Expendable Indian

In 1992, American Indian writer Simon J. Ortiz published *Woven Stone*, a collection of three previously published volumes of poetry—*Going for Rain* (1976), *A Good Journey* (1977), and *Fight Back: For the Sake of the People, For the Sake of the Land* (1980). This third volume most fully voices the ecological problems that have become inescapable for Native peoples facing environmental ruin not of their own making.

Fight Back, his most politicized work, sustains Ortiz's oral narrative style, but is more autobiographical and historical in that it describes the destructive effects of the uranium industry of the 1960s on the people and the land of the Acoma Pueblo community where Ortiz grew up. Ortiz states, "Its stories and poems, although

1960s on the people and the land of the Acoma Pueblo community where Ortiz grew up. Ortiz states, "Its stories and poems, although not written until twenty years later, as the industry was winding down, were being formed in my experience and perception of it in my early adulthood" (22–23). It is in this collection that Ortiz provides an historical account of the exploitation of the people and the land along the Grants, New Mexico, mineral belt and most fully voices his ecological, political, and economic concerns for his people and for all living beings. In *Fight Back,* Ortiz chronicles the various injustices shared by the American Indian community in Northwest New Mexico. His poetry attempts to reproach the blind advancement of modern materialism and technology, expose American governmental hypocrisy, and urge action against environmental and social injustices.

Ortiz opens the first section, "Too Many Sacrifices," with a brief autobiographical sketch discussing how his father performed exhausting work for the Sante Fe Railroad Company for approximately twenty years and warned his sons against a similar fate. Although Ortiz never did work for the railroad, he did spend some time working in the uranium mines and mills, which gave him many of the experiences he recounts in *Fight Back* and helped to form his social and political awareness. Many of the poems in this volume "are narratives that explore the ironies, tragedies, and small personal victories of the workers" (Wiget 109).

In a poem entitled "It Was That Indian," Ortiz tells the story of a Navajo man named Martinez who discovered uranium along what is now the Grants mineral belt when he found a green stone near his hogan. Martinez was celebrated, photographed, and praised as the Indian who brought Grants to prosperity (295) until there were problems. Due to the vast industrial expansion of the mines and mills, Ortiz writes in his poem that people complained about chemicals poisoning streams, mine cave-ins, and cancer caused by uranium radiation (296). Martinez was then blamed by the Chamber of Commerce for the negative side effects of the industry because he found the piece of uranium. Used as a scapegoat by the government, Martinez became the perpetrator of all the harmful effects caused by the government's use of the substance and all the serious destructive consequences that resulted.

the 1961 labor strike so that unions couldn't form, and so that the production of yellowcake (a bright yellow compound produced during the uranium extraction process and used to make nuclear fuel) for the Atomic Energy Commission would not be hindered.

When the American Indians went to the mines for jobs, they were usually employed in low-ranking, dangerous positions because they had no mining experience, as depicted in the poem called "Starting at the Bottom." Ortiz writes that some of the Acoma and Laguna men worked there nearly thirty years at entry-level positions (298). Encountering severe racism, without education or social programs for support, American Indians at the bottom of the hiring hierarchy would stay there, making minimal money. The Natives that remained in such low-ranking jobs were there not only because advanced positions were reserved for whites but also because these jobs were extremely dangerous and the prevailing racist attitude was that Natives were expendable.

"Ray's Story" is a poem about a Muskogee Indian named Lacey who had the hazardous job of pulling from the ore pieces of dynamite, steel, and cable as it came down a chute into the Primary Crusher. Lacey, ordered not to turn off the crusher because it would slow down production, became entangled in a cable, which pulled him through the crusher, where he was instantly killed. The foreman, angry the conveyor belt had stopped running, assumed Lacey was asleep and went to the pit. Seeing Lacey crushed up in the machine, he only said, "Gawd, that Indian was big" (303). Lacey was ultimately blamed for failing to turn off the crusher before removing the cable (303). We are to understand that the foreman's reaction towards Lacey's death is one of cold, dehumanizing indifference. The greatest benefit in keeping the Indians in hazardous jobs was that their deaths could easily be explained and accounted for by directing the responsibility for the death onto the Indian. No one questioned the negligence of an Indian. It was expected. The death of a white man was something the industry had to explain.

Why did the Indians work in the mines? Devastating to the Native way of life, Western expansion resulted in the disruption of self-reliance. Having relied on the land for food and shelter, Native cultures underwent dramatic changes due to European economic encroachment and governmental theft of Native lands for various uses such as railroad production. Native ways of life and land-self-reliance. Having relied on the land for food and shelter, Native cultures underwent dramatic changes due to European economic encroachment and governmental theft of Native lands for various uses such as railroad production. Native ways of life and land-based survival methods conflicted with government laws and ideas of civilization, which is why it became necessary for Natives to work in the mines. In "Final Solution: Jobs, Leaving," Ortiz tells how families were torn apart because fathers had to move away to find work. The pressures to conform to the dominant economic system and to sacrifice their own way of life were daunting. By being forced to participate in the Western cultural economy, families were broken up and experienced new hardships they did not know how to confront. In suffering literal physical removals, either onto reservations or to find work, the Native communities who were devoted to and intrinsically connected to place underwent widespread cultural disintegration.

Throughout *Fight Back*, Ortiz voices his impassioned ecological concerns and describes how industrial and technological advances have affected his homeland and the Native people that inhabited the land. In an interview with Kathleen Manley and Paul W. Rea, Ortiz states that to Native cultures, the "land is a material reality as well as a philosophical, metaphysical idea or concept; land is who we are, land is our identity, land is home place, land is sacred" (365). This Native concept of earth and land as identity, as Ortiz mentions, presents a belief that nature is united with human life, and that it is the sense of place which defines a person's life. Understanding the bond between identity and the land is crucial to understanding nature's significance for Native cultures.

A related ecological theme that appears in *Fight Back* is the concept of balance in the natural world as a means of physical and spiritual survival and cultural continuity. All food sources had to be acquired in such a way as to continue the cycle of giving with the spirit world. For everything taken from the earth, something had to be offered back in return, usually in the form of prayer, ritual or ceremony. Neglecting to perpetuate this cycle of giving could result in the permanent loss of something, such as species extinction, disrupting the necessary balance upon which the natural world thrives and regenerates. When this balance is physical, and spiritual well-being rests on conditions of landscape and harmony within the natural world. The earth is regarded as sacred space, differing from Western perspectives that view land as real estate to be bought and sold in a commodity market.

In "Returning It Back, You Will Go On," Ortiz writes of how technological advances (power companies, railroads, mining corporations, etc.) have encroached on Indian lands and stolen from the earth. Ortiz believes that the land will only be able to renew itself if America learns to makes its own offerings in return (331). Reverence and respect for the earth entail a cycle of giving or the earth will not give again the natural resources that were stripped from it. Once the land is taken, so is the identity and power of the American Indians. Ortiz claims that if the environment is not nurtured, the survival of the Native people is at risk, and the land will ultimately suffer depletion (331). An ecological consciousness with a sense of reverence then, is necessary for environmental and cultural continuity.

Ortiz first begins his criticism of land exploitation in *A Good Journey* with the poem "A Designated National Park," which reproduces a sign outside of Montezuma Castle, Arizona, that requires a certain fee in order to enter the area. In order to go home, Ortiz writes, that he has "to buy a permit" (235). The construction of boundaries in order to claim ownership for the purpose of commercialization is one of the major criticisms that Ortiz has concerning Western land use. According to William Cronon, "more than anything else, it was the treatment of land and property as commodities traded at market that distinguished English conceptions of ownership from Indian ones" (75). Tribes occupied certain territories, but occupation was not ownership. The landscape with its spiritual significance was considered a living entity beyond the scope of human claim or monetary value. That Europeans were parceling out patches of land and selling them for a price was seen as disrespectful and arrogant.

Perhaps one of the most disturbing aspects of Ortiz's account in *Woven Stone* is the European ethnocentrism that justified exploitation of the land and betrayal of the Acoma people. When the first Spaniard entered the lands in 1540, "he recommended occupation and settlement because of the natural material wealth of the land . . . the land and the people were obviously productive and the potential for colonization and profit was worthy of royal

Perhaps one of the most disturbing aspects of Ortiz's account in *Woven Stone* is the European ethnocentrism that justified exploitation of the land and betrayal of the Acoma people. When the first Spaniard entered the lands in 1540, "he recommended occupation and settlement because of the natural material wealth of the land . . . the land and the people were obviously productive and the potential for colonization and profit was worthy of royal and private investment" (341–342). Without any respect or consideration for the people who inhabited the land when they arrived, the Spanish were primarily concerned with its monetary possibilities and claimed the land as their own. Their heavy development left the once lush and grassy area barren, dry, and brittle. Ortiz writes: "The railroads were the first large Industrial users of the water belonging to the land and people. They found it easy enough to get it; they simply took it" (343).

The vast changes that took place for the Acoma during this time often left them frustrated and deeply concerned for their future. It was easy for the whites to steal land from them when they couldn't read English documents or communicate effectively with the whites. Up until that time, the people were able to survive on their own by agricultural means without earning money or living in European-style homes. But once colonization was initiated, the Natives could not survive and carry out their lives on their own terms. According to Ortiz, "The people had always been able to deal with the earth, even its barren times, even its rainless times, on their own terms. But when the Mericano system caused dependency, the people became bewildered and often helpless" (351). Forced to adopt foreign ways that went against their cultural beliefs, the Acomas were susceptible to European domination and betrayal. Their lands lost their native names, and the people were often dislocated to reservations or lands which were unproductive and thus of no economic use to the government.

Ortiz discusses American development in terms of its social and economic hypocrisy: "The nation swept on into the 20th century and the Mericano was not called thief or killer; instead he was a missionary, merchant and businessman, philanthropist, educator, civil servant, and worker" (350). Ortiz claims that individuals practicing the capitalist American endeavor, or Manifest Destiny, were regarded as heroes or people making government to maintain control over them by, for example, using relatively unpopulated Indian lands for nuclear test sites. And the hypocrisy still survives:

> The American poor and the workers and white middleclass, who are probably the most ignorant of all U.S. citizens, must understand how they, like Indian people, are forced to serve a national interest, controlled by capitalist vested interests in collusion with U.S. policy markers, which does not serve them. (361)

The concern here is not just for American Indian peoples, but for all Americans who are robbed by a materialist system. His activism becomes more effective as he opens up the scope of victimization to include those from Western cultural backgrounds who may have been alienated by his prior concerns. Near the end of "Our Homeland, A National Sacrifice Area," Ortiz voices the danger that awaits all Americans if their lives keep racing destructively in the same direction: "If the survival and quality of the life of the Indian peoples is not assured, then no one else's life is, because those same economic, social, and political forces which destroy them will surely destroy others" (360). Once again, Ortiz reduces the gap between Native cultures and Western cultures by emphasizing the human plane we all inhabit and the destructive forces to which we are all vulnerable.

In one interview, Ortiz expresses his concerns for the future of the human race in the wake of an all-consuming industrialism: "I fight against technological industrialism simply as a human being, because I am a human being and I want to remain human, very close to what human emotions are, very close to what the human spirit is, and hopefully to convey this through literature" (Manley and Rae 370). Ortiz, fighting against industry, hypocrisy, and landscape devastation in *Fight Back*, is opposed to the dehumanization of all of us who are either consciously or unconsciously suffering from modern-day alienation from what is naturally human. Ortiz knows that that which goes against being human estranges people from each other and themselves as well as from the nonhuman natural world.

In *Fight Back: For the Sake of The People, For the Sake of the Land,* Simon Ortiz, through his highly accessible narrative voice, tells the story of his heritage in the Acoma pueblo culture and landscape devastation in *Fight Back*, is opposed to the dehumanization of all of us who are either consciously or unconsciously suffering from modern-day alienation from what is naturally human. Ortiz knows that that which goes against being human estranges people from each other and themselves as well as from the nonhuman natural world.

In *Fight Back: For the Sake of The People, For the Sake of the Land,* Simon Ortiz, through his highly accessible narrative voice, tells the story of his heritage in the Acoma pueblo culture and keeps the memory of Acoma's experiences with the white civilization alive. In this volume, Ortiz truly is fighting back—against oppression, colonization, exploitation, and technological progress that is blind, destructive, and dehumanizing. By relaying these stories, Ortiz contextualizes American capitalist injustice and hypocrisy in order to set the stage for change, which begins with redefining progress as we saw in Susan Griffin's work. In the interview with Manley and Rae, Ortiz asserts that if technology goes against being human, then we have to fight and resist it, and that we need to pursue a creative relationship with technology in order to avoid becoming complacently dependent upon it (371). Ortiz's activism resonates in his attempt to reterritorialize how we view technology as a symbol of progress and goodness in order to take a long critical look at its devastating side effects. Although Ortiz knows he cannot stop technology, his commitment to that which is human remains unwavering. While he resists it, he reterritorializes technology to use it for a source of creative power, a power which he has eloquently used in writing *Fight Back* to demand that "No More Sacrifices" be made that prove unjust to the land and the people who inhabit it.

Wendy Rose: Anthropological Activism

Wendy Rose is an American Indian poet whose Hopi/Miwok mixed ancestry plays an important role in the development of her creative writing voice. For Rose, this "half-breed" status is critical because the Hopi culture is matrilineal, only recognizing those with a Hopi mother as part of the tribe. Rose's mother was Miwok and her father was Hopi, but she still identifies herself as Hopi,

Acoma pueblo area in the southwestern United States. For Rose, home is Coursegold, California, near Fresno.

Wendy Rose's poetry presents an activism that is academic, environmental, and cultural. In addition to being a poet, Rose has a degree in anthropology. Much American Indian literature and various postcolonial theories address the need for colonized cultures to take control over the body of knowledge about their lives and histories because this information has been produced by outsiders with Eurocentric and even racist posturings who misinform and misrepresent in accordance with those biases. In an interview with Laura Coltelli, Rose calls herself a spy who wants to help control the handling of artifacts and attitudes of other anthropologists (*Winged Words* 124). This form of activism is clearly present in many of her poems that address the issues of dehumanization and disrespect involved in the anthropological excavation of American Indian artifacts from their homelands.

Trinh T. Minh-ha in *Woman, Native, Other* (1989) provides a valuable discussion of the subjectivity of the anthropological science. For Trinh, anthropology is not a hard, factual science but more like "scientific gossip." She criticizes anthropological interpretation as an institutionalized means of seeking a stereotype of the Other while the field and its practitioners refuse to examine their own position within a community of colonizers.

Minh-ha also describes anthropologic writings as being equal to fiction: "The dilemma lies in that fact that descriptions of native life, although not necessarily false or unfactual, are 'actor-oriented,' that is to say, reconstructed or fashioned according to an individual's imagination" (70). This statement is similar to Hayden White's point of view concerning the fictionality of the historical text discussed in chapter two. The "individual's imagination" in Minh-ha's statement is the imagination of the anthropologist who feels obligated to "interpret" a Native's account, which is assumed to be inaccurate or inconclusive. This places the anthropologist in the position of authority to intrude upon Native peoples or to manipulate and distort ethnographic knowledge in the name of hard science or academic advancement. Rose's grave distrust of these disciplines suggests that the misrepresentation and ideological prejudice practiced by anthropologists are often willful and fully realized. Her endeavors in anthropology and in her poetry are aimed at reclaiming some historical text discussed in chapter two. The "individual's imagination" in Minh-ha's statement is the imagination of the anthropologist who feels obligated to "interpret" a Native's account, which is assumed to be inaccurate or inconclusive. This places the anthropologist in the position of authority to intrude upon Native peoples or to manipulate and distort ethnographic knowledge in the name of hard science or academic advancement. Rose's grave distrust of these disciplines suggests that the misrepresentation and ideological prejudice practiced by anthropologists are often willful and fully realized. Her endeavors in anthropology and in her poetry are aimed at reclaiming some authority and changing the attitudes of scholars and students in the fields of anthropology/archaeology who continue to disrupt sacred burial grounds and reduce Native cultures to "objects" of scientific analysis.

Although Rose claims to be an "urbanized Indian," her poetic voice is unmistakably connected to the natural world. The first poem in *Bone Dance: New and Selected Poems, 1965–1993* that addresses the issue of the mistreatment of American Indian artifacts and links such treatment to an orientation toward the natural world is "Lab Genesis." This poem describes survival despite the calcification of American Indian cultures in the hands of archaeological science. Rose states explicitly in the poem "there will be / no archaeology / to my bones" (6). Images throughout the poem stress the continuity of life after death and the inability of any science to remove her life from the natural world:

> life is dying
> each moment
> learning to live
> each moment
> in & out
> like bird breath
> like toad's tongue
> like making love (6)

Rose equates death with learning to live but in a different form. She also emphasizes the natural cyclical processes of other forms of life. This is reminiscent of the way that Wendell Berry addresses death. Rose suggests cremation at the end of the poem to prevent any future desecration of her bones.

In the interview with Laura Coltelli mentioned earlier, Rose states that her desire to protect Indian burial grounds is both metaphorical and literal: "The metaphor is to protect Indian people through, in some instances, trying to neutralize the very weapons that are being used against Indians, by mastering those weapons and then in a sense breaking them from within" (125). Her activism is literal as well in that one poem she has written was composed in San Jose, "in front of a bulldozer, on top of an Indian cemetery, where [she was] sitting to prevent the bulldozer from just going through and ripping up the Indian graves" (125). Rose attempts to dismantle the colonialist ideology by using the master's tools in hopes that the change in attitude will be followed by a change in behavior. But in the meantime, she and others put themselves in harm's way in order to protect what is sacred and meaningful in the landscape. According to the interview, this particular attempt at stopping the destruction of the burial ground in San Jose was successful. The land and its inhabitants were preserved.

One of the most well-known poems by Rose is the widely anthologized "The Three Thousand Dollar Death Song." This poem begins with an italicized quote from a museum invoice:

> *Nineteen American Indian skeletons from Nevada . . .*
> *valued at $3,000.*
>
> —*invoice received at a museum*
> *as normal business, 1975*

Then Rose's poetic voice begins:

> Invoiced now
> It's official how our bones are valued
> that stretch out pointing to sunrise
> or are flexed into one last fetal bend,
> they are removed and tossed about,
> catalogued, numbered with black ink
> on newly-white foreheads. (20)

Rose's profound discontent is unclouded as she describes the dehumanization of her ancestors' bones. These bones to Rose are priceless; they are the remains of family members. Her outrage stems from the idea that Western scientific and academic practice assumes an imperial posturing and places a price on these bones, reducing them to a marketable commodity. The bones are also handled by strangers and then objectified as materials for scientific analysis. For cultures that have suffered deaths of holocaustal proportions, oral histories perished with the people, and these bones are sacred as perhaps the only remaining proof of a once-thriving civilization. They are connections to the spirit world where ancestors are considered still very much alive. In this poem, Rose identifies the degrading effect archaeological science has on these sacred artifacts:

> one single century has turned
> our dead into specimens,
> our history into dust,
> our survivors into clowns. (21)

Archaeological science is a further obstacle to Native cultures attempting to maintain the history of their own people without the corruption of Western culture science and capitalism.

In the above passage, Rose also mentions that in the last century the survivors have been turned into "clowns." A mid-nineteenth century book entitled *History of the Indian Tribes of North America* by Thomas L. McKenney and James Hall, predicted the disappearance of North American Indian tribes. The book includes portraits of American Indians, primarily from the Great Lakes region. The portraits were reprinted in a publication by the Library of Congress called *North American Indian Portfolios* along with paintings by George Catlin and Karl Bodmer. These portraits unfortunately meet Rose's clown-like description. Most of the portraits show Natives wearing odd mixtures of brightly colored colonial jackets over ruffled high-neck shirts in conjunction with their tribal headdresses. The tribal facial markings look false and gaudy, as the expressionless Natives are pose rigidly wearing silver amulets around their necks that bear President Jackson's portrait.

James Gilreath, an American history specialist in the Rare Books and Special Collections department of the Library of Congress, states that the portraits by McKenney and Hall are "stiff and wooden." He also states that they are "exotic, unknowable, and perhaps a little frightening. . . . The coloring on each Indian differs, but the poses are almost exactly the same, and no personality is shown. They are less individuals than parts of a species to be observed" (9). Rose would agree. These portraits tell less about the Indians of the nineteenth century and more about white cultural attitudes toward Natives that have prevailed into the twentieth century.

Two other poems in *Bone Dance* that address the desecration of Native bones are "Excavation at Santa Barbara Mission" and "Retrieving Osceola's Head." "Excavation at Santa Barbara Mission" tells of archaeologists who discovered that the Spaniards built the mission from the bones of dead Indians. In this poem, Rose speaks of herself as one of the archaeologists "crouching in white dust, / listening to the whistle / of longbones breaking / apart like memories" (85). Throughout the poem, Rose speaks of discovering "Marrow / like lace" and "fingerbones / scattered like corn / and ribs interlaced / like cholla" (84). Her shock and horror echoes at the end of the poem with this repetition:

> They built the mission with dead Indians.
> They built the mission with dead Indians.
> They built the mission with dead Indians.
> They built the mission with dead Indians. (85)

"Retrieving Osceola's Head" is about a scientist who decapitated Osceola, a famous Seminole chief, and took the head home with him. Then he used it to punish his children by hanging it on their bedpost at night when they went to bed. Early in the poem, Rose writes:

> but we have learned
> to keep our heads
> and never forget
> how it was
> in the grave. (106)

Native cultures are constantly fighting battles to protect what is rightfully theirs whether it is in the form of land or sacred artifacts. They have learned to keep careful watch over the bones of their ancestors, and activists like Rose, are attempting to break into the academy to help control the system of knowledge. The ending of the poem is an expression of transformation and empowerment:

> We became peaches
> bursting through
> the end of summer.
> We became
> all of the murders
> returning
> We
> became whole again. (107)

The deaths come back as life in a different form, and those who were torn from their people and their homes become whole again.

Chrystos, also a American Indian poet, wrote a scathing and humorous criticism of the fields of archaeology and anthropology in a poem appropriately entitled "Anthropology" published in *Dream On* (1991). In "Anthropology," Chrystos adopts the voice of an anthropologist from an assumed superior culture who is analyzing the habits and practices of the Caucasian culture, which she refers to as "cauks for ease in translation" (78). Her ethnocentric posturing in the poem is unmistakable and mirrors the way that white anthropologists have viewed American Indian cultures from their self-imposed position of authority.

Her descriptions of the "cauks" make them look absurd as she turns the tables and makes the white culture the object of analysis: "The most important religious ritual, one central to all groups, is the / mixing of feces & urine with water. This rite occurs regularly on a / daily basis & seems to be a cornerstone of the culture's belief system" (78). It is not an accident that Chrystos targets aspects of the white culture that anthropologists have criticized in Native cultures, such as religious practice and land use: "The main function of the majority of non-city dwellers is the / production of an object called a lawn." This lawn appears "to / have a sacred character, as no activity occurs on it & keeping it short / green &

square is a constant activity" (78). Chrystos makes no mistake about speaking from a deliberate American Indian cultural bias in the same way that Native cultures have been reduced to objects by the Western cultural examination and zoo-effect observations of Caucasian anthropologists. American Indian philosophies did not include ownership of a plot of land to be groomed to a strict aesthetic uniformity. What is the purpose of that ritual? Chrystos wants anthropologists to know that the European cultures have practices that look absurd and bizarre to other cultures as well. One of the criticisms of anthropology made by Trinh Minh-ha, as discussed earlier, has been directed toward the field's practitioners and their inability to self-examine and recognize their equal footing among the human race, complete with their own collection of biased practices and worldviews.

She continues the poem poking fun at the "cauks" and their ritual of getting food from "pushbutton machines or / orange plastic small markets" where her researchers found the food to be "completely inedible" (78). Her criticism of Caucasians escalates when she writes of animals, children, the elderly, and people who are physically or mentally ill being jailed and that "the actual spiritual purpose of the culture, is to jail as much as possible. Extensive use of fences is the key argument for this theory" (79). She closes the poem with a criticism of how anthropologists exploit American Indian remains in the name of science: "Our data is as yet incomplete. We hope by 1992 to have a more comprehensive overview, at which time a traveling exhibition of artifacts (including exhumed bodies to illustrate their burial practices) will tour for the education of all" (79). Wendy Rose would read this poem with a smile of approval. "Anthropology" humorously and very seriously exposes the unconscious cultural superiority that Western culture has assumed over American Indians.

Wendy Rose's poetry is often angry and resentful. More than any other American Indian poet, she seems distressed by the inhumane treatment of Native bodies and materials by archaeologists and anthropologists. Her passion for change is proven by her commitment to acquire a Ph.D. in anthropology in order to affect the biased system from within. Her efforts to gain control of Indian artifacts have not been entirely successful.

Gerald Vizenor is arguably even more offended by anthropological methodologies than Rose. In an interview with Laura Coltelli from *Winged Words*, Vizenor agrees that "everything in anthropology is an invention and an extension of the cultural colonialism of Western expansion" (Coltelli 161). He views anthropology as an academic and material creation formulated to allow Western academics to continue their illusion of being in control of other cultures when they're realistically only inventing cultures in order to progress in universities. In the same interview with Coltelli, he explains that there is nothing real or authentic about anthropology, only stories and symbols that have been either borrowed or fabricated (161–163).

His *The People Named the Chippewa* (1984) presents Chippewa history without ethnological or anthropological invention. Vizenor reviews tribal and reservation publications, and radio and television stations that attempt to arrest some of the media formations from the white culture and how these efforts have been continuously thwarted. The prevailing message is that information presented by the white culture is not to be trusted. He writes: "The cultural and political histories of the Anishinaabeg were written in a colonial language by those who invented the Indian, renamed the tribes, allotted the land, divided ancestries by geometric degrees of blood, and categorized identities on federal reservations" (19). Although Vizenor might commend Rose for her desire to challenge biased anthropological science from within the academy, his activism confronts dehumanization, environmental degradation, and anthropological constructions using a different strategy—the trickster.

The trickster figure is one of the richest aspects of American Indian oral traditions. Having many functions—a teacher, a form of constructive chaos, or an administrator of tribal jokes and pranks—the trickster figure is vital to American Indian continuance and provides one of the best forms of healing for a fragmented culture—humor. At times the trickster's lessons are deadly serious, and at times they are quite silly. Either way, he (usually male) always plays a vital role in the Native oral tradition. Some of the trickster's functions are to teach the ability to laugh at one's self, to act as a vehicle for unacceptable behavior to reduce tribal misconduct, and to demonstrate the importance of transformation. The trickster figure is very malleable, taking on many forms both human and animal. Most often, the trickster comes in the form of a crow, rabbit, coyote, or old man. His unpredictability and transformative abilities are what give him such power in teaching his wisdom. Alan R. Velie in his article "Gerald Vizenor's Indian Gothic," comments on the validity of the trickster figure as a form of organized chaos and as an outlet for unacceptable behavior:

> Any society that has oppressive rules of moral and ceremonial behavior needs mythic and ritual sources of rebellion which allow tribal members to flout the rules through surrogates. The surrogates were irresponsible, amoral figures who mocked everything sacred with impunity to the delight of the rest of the community which remained obedient and orderly. (81)

The trickster, therefore, acted out immoral impulses or temptations for the tribal members in order to reduce any desire they may have to carry out the act in reality and break tribal codes of conduct that may lead to ostracization.

Gerald Vizenor, a trickster himself, tricks his readers into learning about the dangers of terminal creeds and ecological exploitation in his first novel *Bearheart: The Heirship Chronicles*, by shocking them with explicit and brutal sex and violence. In the novel, Vizenor depicts a postapocalyptic wasteland of nearly total ecological destruction. The only remaining place untainted by the modern technological forces is the Cedarfair forest, where it is only a matter of time before the cedar nation is felled by the federal government's greed. According to Alan R. Velie, Vizenor reverses the values of the ideological constructs of the frontier, which defined the forest as evil and civilization as good. In *Bearheart* "the forest is good, a source of strength, and civilization is evil, a corruptive influence on man" (84). By reversing historical and ideological constructions of the wilderness into a place of health and balance, Vizenor relocates sources of evil and corruption away from the forest and into places of development to expose who and what pose real dangers to human survival.

Vizenor's futuristic wasteland lies on the border between reality and the absurd, making it easy to contemplate his stark vision as a real possibility. The wasteland exists because "the nation ran out of gasoline and fuel oil. Electrical power generating plants closed down. Cities were gasless and dark" (23). Vizenor plays upon the fears of his readers who know that if the land and our energy forms continue to be consumed at alarming rates, the resources will be gone. Many of the characters are victims of technological violence, "terminal creeds," selfishness, greed and environmental exploitation, which as the trickster teaches us, lie at the core of societal corruption.

Throughout the novel, a group of traveling pilgrims on their journey to the fourth world, come across characters who are victims of the technological violence that has destroyed the earth's balance. Proude Cedarfair is the main trickster figure who beats Sir Cecil Staples at his own game through his wisdom and ethics. The trickster figure in this book is the one who survives despite all the obstacles. Because he is such an integral part of the oral tradition and embodies the beliefs and values of Native attitudes toward transformation and environmental respect, he proceeds to the fourth world. Vizenor writes that "the third world turns evil with contempt for living and fear of death" while "in the fourth world evil spirits are outwitted in the secret languages of animals and birds" (5). *Bearheart* is a depiction of the third world in which characters who continue to exploit the earth or who embrace terminal creeds will perish.

The most significant of these characters is Sir Cecil Staples, the evil gambler, who gambles with motorists for gasoline. If they lose, which all do except for Proude, they get to choose how they will die. He is the epitome of evil, darkness, and all that is vile and corrupt. The pilgrims on their visit to the gambler for gasoline, experience the gambler's evil immediately: "Proude Cedarfair could smell ammonia when he entered the trailer. . . . Death has the smell of cities and machines and plastics. There was death and evil in the altar trailer" (120). It is not until Sir Cecil begins talking about his childhood that it becomes clear that he is the personification of the dangers of technological advancement. His mother's apparent disgust for insects motivated her to kill them obsessively with poisonous insecticides. Sir Cecil states: "Dear mother, she poisoned us all, but she did kill what she wanted to kill. But we inhaled insect poisons all the time we were living together" (123).

The evil gambler's squeamish mother not only poisoned her son physically but also turned him into a murderer: "Torture and death. Not the killing of insects like my mother but the death of real people, is an obsession with me"(123). Later, the evil gambler does not attribute his violent tendencies to his mother but rather to the government and big businesses that "started [the] indifference toward death with their pollution and industrial poisons" (127). Sir Cecil's fascination with death and killing is made more horrific by his pleasure in torturing people. This, he states, is also a result of modern technology. Sir Cecil Staples is the embodiment of the problems of a toxic, polluted living world and other repressed evils that have been repressed and improperly dealt with, such as child abuse and violent crime.

Along with Vizenor's ecological theme questioning the blind progression of technology is his ideological activism. Several of Vizenor's characters have "terminal creeds" which lead the characters to their untimely deaths. Proude Cedarfair confronts and undermines the power of terminal creeds, which are produced by academics and government officials who wish to inscribe tribal peoples with a static identity. It is a means of maintaining control over Native knowledge and substantiating a preexisting ethnocentrism. Terminal creeds refer to ideologies and meanings that are fixed and left unchallenged.

When the pilgrims enter the bioavaricious Regional Word Hospital, Vizenor plays upon the slippage between the sign and the signifier in order to expose the fallacy of fixed meanings in language. The staff at the word hospital supports attempts to mechanize and analyze word meanings in order to prevent future misunderstandings in communication. One hospital attendant claims that "the breakdown in law and order, the desecration of institutions, the hardhearted investigations, but most of all the breakdown in traditional families was a breakdown in communication" (166). So, at the word hospital, they attempt to fix broken words into static meanings to prevent future breakdowns. Hospital workers are sitting at computers entering secret, classified words and phrases into a computer to undergo a color coding indicating value and chromatic meaning. It all sounds ridiculous, but Vizenor contrasts this word hospital of terminal language with the oral tradition, which values the imagination and the performative telling of the story over establishing one definable meaning. In the oral tradition, the story changes spontaneously with each teller, and fluid symbolic meanings take precedence over a story that claims a single meaning or objective truth. Vizenor writes that, "survival on the interstate was more verbal than spiritual, words not silence, more open than closed, less secret, little political" (161). Those who knew the power of language were the strongest survivors. Vizenor's objection relates to the continued effort by Western cultures to turn everything human into a science. In describing the atmosphere at the word hospital Vizenor writes, "The machines were humanized while the humans were mechanized" (167). Like Ortiz and Rose, Vizenor continues to identify ways in which the modern, industrialized, scientific standards of living have devalued human life.

In *Bearheart*, one significant terminal creed is the definition of an Indian. In the chapter "Terminal Creeds at Orion," Vizenor launches into an amusing yet important scene where Belladonna Darwin-Winter Catcher finds herself stumbling through an interrogation by the hunters of Orion who want to know what an Indian is. When asked to describe what an Indian is, Belladonna provides a lengthy list of clichés. Some of these clichés include: "Indians have more magic in their lives than white people," and "Indians have their religion in common" (195). When asked again, she repeats the stereotypes, indicating that she is a victim of Western cultural invention of the Indian and a person who accepts what information she receives without question.

Through Belladonna, Vizenor deconstructs the definition that Western anthropologists have given to the Indian. The hunters accuse Belladonna of voicing "collective generalizations" and of not speaking as "a person of real experience and critical substance" (196).

In agreement, Louis Owens in his "Afterword" to the novel remarks on Belladonna's performance: "Speaking as a romantic invention indeed, a reductionist definition of being that would deny possibilities of the life-giving change and adaptation at the center of traditional tribal identity" (280). Because the hunters believe Belladonna is suffering from the illness of terminal creeds, they kill her by feeding her sugar cookies laced with alkaloid poison—"The poison cookie was the special dessert for narcissists and believers in terminal creeds. She was her own victim" (199). She was her own victim, but she was also a victim of Western cultural assimilation and invention.

Belladonna, when asked what it means to be Indian, responded to the stereotyped definitions that attempt to homogenize and classify diverse cultures. This is not to suggest that locating and understanding one's identity as an American Indian would be an easy journey. Much contemporary American Indian literature portrays a character in search of his/her tribal identity, which has often been distantly removed from its place of origin and heavily diluted by the Western values and capitalism. But, as Vizenor the trickster teaches with the demise of Belladonna, it is a necessary journey to make.

As a postcolonial text that examines the relationship between center and periphery and helps to destabilize colonial constructs, *Bearheart* is an acerbic criticism of Western philosophy and its buried corruptions. Interestingly, the novel has also been hailed as postmodern because of its shock value, elements of chance, indeterminacy, and deconstructionist tendencies. As mentioned in the introduction, postcolonial critics Vihay Mishra and Bob Hodge assert that a postcolonial text that is also postmodern is ineffective because it ultimately subverts the postcolonial agenda, that by taking on a master narrative literary form or distinction it has lost its power. But isn't it true that much postcolonial literature demonstrates its political struggle by using the master's tools to dismantle the house? What critics also may not realize is that elements of *Bearheart* that have been assigned postmodern value existed in the American Indian oral tradition, particularly in trickster tales, long before postmodernism was launched into academic discourse. Chance, vulgarity, play, coincidence, and transformation are all characteristic of oral tradition storytelling. Perhaps postmodernism is the Western culture finally recognizing its own form of tricksterism as its values and practices become increasingly fragmented and consistently challenged.

As the main trickster in *Bearheart*, Gerald Vizenor is intelligent, thorough, and relentless. He unapologetically illustrates in his novel that conquering nature is an illusion, that terminal creeds will provide no personal growth, that academics and anthropologists have invented a stereotyped Indian, and that the Western world has repressed problems that need to be addressed. In doing so, Vizenor revokes ideologies that will lead to human and ecological destruction. His appeal is to fear by presenting to his readers a world void of human respect and dignity. Vizenor, the trickster, knows that a change in one's behavior must be preceded by a change in one's way of thinking and that fear, used ethically, can be a persuasive tactic. Proude Cedarfair, because of his spirituality and strong moral character, is the only one of the primary characters to make it to the fourth world, and this is at the heart of Vizenor's ideological activism. That Proude turns into a bear before entering the fourth world is important as well in reestablishing the significance of the oral tradition and the worldview it fosters.

Simon Ortiz, Wendy Rose, and Gerald Vizenor have in common an activism that fights against dehumanization of the earth and American Indian cultures. They document the ways in which indigenous peoples have been historically exploited throughout their lives and even after death. By investigating the concept of reterritorialization through American Indian activist writers, the process becomes apparent as a political act, one that is nonprescriptive, demanding cultural specificity and an agency for change. At the same time, reterritorialization is a personal matter. Each writer has his or her own area or subject of emphasis which each feels is in need of reconsideration. Issues surrounding technology, artifact excavation, and fixed ideologies are foregrounded in these writers but always pivoting around an environmental center. A disruption of the earth/human balance is at the heart of all illness, suffering, and social disorder. Native perspectives teach that if we interact with the nonhuman living world with violence and disrespect, then that violence will circle back toward human beings. Environmental concerns are central to the issues Ortiz, Rose, and Vizenor address because the healing of the earth/human balance is central to healing all other forms of injustice.

Chapter Six

Concluding Remarks

> The concept of country, homeland, dwelling place becomes simplified as "the environment"—that is, what surrounds us. Once we see our place, our part of the world, as surrounding us, we have already made a profound division between it and ourselves. We have given up the understanding—dropped it out of our language and so out of our thought—that we and our country create one another, depend upon one another, are literally part of one another; that our land passes in and out of our bodies just as our bodies pass in and out of our land; that as we and our land are part of one another, so all who are living as neighbors here, human and plant and animal, are part of one another, and so cannot possibly flourish alone; that, therefore, our culture must be our response to our place, our culture and our place are images of each other and inseparable from each other, and so neither can be better than the other.
>
> —Wendell Berry, *The Unsettling of America*

As mentioned in the introduction, the impetus for this study of reterritorialization came from the article by Caren Kaplan called "Deterritorializations: The Rewriting of Home and Exile in Western Feminist Discourse." Kaplan focuses primarily on autobiography in her essay, while this project includes works of fiction, poetry, creative nonfiction, and cross-genre works, operating under the philosophy that home and a sense of self are sought through writing in multiple genres. The quest for connection between self, community, history, and self is by no means restricted to memoirs. The quest stems from a basic human need for safety and acceptance, which can come from any writer, of any ethnic origin, who is sensible to the feeling of not having a safe place to inhabit either physically or psychologically.

Featured prominently in this exploration are American multicultural writers who have experienced a colonizing form of displacement. Most of the authors are of American Indian ancestry. The literature by these authors works particularly well for this analysis due to the historical removals Native cultures have undergone in the European colonizing process and also because of the centrality of the landscape and sense of place in many Native philosophies and practices.

The application of the concept of reterritorialization to American Indian writings entails working against romantic and idealized images of Natives as people already in tune with nature through recognition that inhabitation requires effort on their part as well. Posters and greeting cards at any local New Age store show the stereotyped noble Indians in full Indian dress staring out over the western landscape as their long hair blows on the windy plain. These romantic images not only calcify and commodify a cultural image but also undermine the displacements Indian peoples have undergone on the American continents.

Part of the reterritorialization effort involves recovery of lost stories and cultural practices, but that effort also involves imagination and invention. Most of the landscapes that the ancestors of these writers inhabited are strongly altered by technology and progress so that a return to place could only be metaphorical or imaginative. There also has been so much assimilation and cultural exchange that a return to cultural purity would be impossible. Reterritorialization, then, can act as a decolonization process that is often highly nostalgic and spiritual. Still, the process or outcome is no less authentic or meaningful. Being in the world is as much a psychological act as a physical one. And for many American Indian cultures, time and space are fluid multidimensional spheres rather than linear planes of existence. Experiences gained through the power of the mind and the imagination are not valued less than physical experience.

In *Grandmothers of the Light*, Paula Gunn Allen speaks of this phenomenon as "the plasticity of time in the universe of power." The universe of power is a place, she describes, where the sources of magic and transformation reside. In this passage she compares the ordinary world and the universe of power:

> In the ordinary world, we get from one place to another by walking, running, riding on animals, or riding or flying in machines. . . . But in the universe of power we, our signals, and our objects can traverse great or small distances with the speed at which a message can presently be sent over a fax machine. Objects and subjects alike can be transported through solid matter—windows, walls, stone buttresses, or mountains—and they are as independent of gravity as of other physical constraints. (17)

The universe of power is a place of much greater agency than the material world. It is a place where painful memories of colonization and betrayal can be transcended. The spiritual and psychological power of this universe enables one to surpass and transform the difficulties of the physical world. Consequently, the universe of power is a great healer.

This conclusion resonates from my point of departure, the discussion of home. As demonstrated by the writers explored, the concept of home involves a series of boundaries and dimensions that are highly individual. These boundaries are particularly evident in Chrystos, who is unreserved about claiming her space and ostracizing unwanted occupants. Home also needs a basis for comparison. Understanding a sense of place invariably involves differentiating that place from others and knowing its particularities. The local culture of a place involves a long history of its people and economic opportunities as well as physical attractions that led to settlement and growth. The peculiarities of language and social practice of place have their roots in the people who initially settled there, and these characteristics persist and evolve as new settlers bring their influence.

Mythic, psychic, and environmental reterritorializations involve the claiming of space for oneself and an understanding of the place's history, its physical constituents, and one's own psychological reaction to these aspects. The writers discussed are in tune with the fluidity of place within the individual as an exterior force that has profound effects on the interior sense of well-being. We are our environments. We take in physically and psychologically our surroundings, and they become part of who we are. That is why it is of great importance for our surroundings to be healthy and habitable. Place and the self are not separate entities, which is why the explorations of this project pivot around an environmental center.

Postcolonial thought, ecocriticism, and theories on the sense of place are all valid approaches to literary studies because they attempt to diffuse assumptions and behaviors that have tenured the Western power paradigm in terms of gender, culture, and capitalist progress. To put it bluntly, the world has its share of challenges. The more these problems escalate, the more there is a need for theories that address different ways of approaching how we live within the world and among one another. A quote that comes from Wendell Berry's *The Unsettling of America* states that "we can make ourselves whole only by accepting our partiality, by living within our limits, by being human—not by trying to be gods" (95). Our existing approaches to being in the world need to be modified, and our values need to be seriously reconsidered. Caring more for the other does not mean caring for the self less. It means recognizing one's position within the multitude of life forms in the universe and taking some responsibility for their well-being. If one is not safe, none of us are.

References

Alaimo, Stacy. "Displacing Darwin and Descartes: The Bodily Transgressions of Fielding Burke, Octavia Butler, and Linda Hogan." *ISLE: Interdisciplinary Studies in Literature and Environment* 3.1 (Summer 1996): 47–66.

Alarcón, Norma. "Chicana Feminism: In the Tracks of 'the' Native Woman." *Understanding Others: Cultural and Cross-Cultural Studies and the Teaching of Literature.* Eds. Joseph Trimmer and Tilly Warnock. Urbana: NCTE, 1992. 96–106.

Allen, Paula Gunn. *Grandmothers of the Light: A Medicine Woman's Sourcebook.* Boston: Beacon, 1991.

———. "The Sacred Hoop: A Contemporary Perspective." *The Sacred Hoop: Recovering the Feminine in American Indian Traditions.* Boston: Beacon, 1986. 54–75.

Anzaldúa, Gloria. *Borderlands/La Frontera: The New Mestiza.* San Francisco: Aunt Lute, 1987.

Appadurai, Arjun. "Disjuncture and Difference in the Global Cultural Economy." Williams and Chrisman 324–338.

Armbruster, Karla. "Blurring Boundaries in Ursula Le Guin's 'Buffalo Gals Won't You Come Out Tonight': A Poststructuralist Approach to Ecofeminist Criticism." *ISLE: Interdisciplinary Studies in Literature and Environment* 3.1 (Summer 1996): 17–46.

Ashcroft, Bill, Gareth Griffiths, and Hellen Tiffin. *The Empire Writes Back: Theory and Practice in Post-Colonial Literatures.* New York: Routledge, 1989.

Austin, Mary. *The Land of Little Rain.* 1903. New York: Dover, 1996.

Bakhtin, M. M. *Toward a Philosophy of the Act.* Trans. Vadim Liapunov. Ed. Liapunov and Michael Holquist. Austin: U of Texas P, 1993.

[Bakhtin]/V. N. Vološinov. *Freudianism: A Marxist Critique.* Trans. I. R. Titunik. Ed. Titunik and Neal H. Bruss. New York: Academic Press, 1976.

Berry, Wendell. *Clearing.* New York: Harcourt Brace, 1974.

———. *The Country of Marriage.* New York: Harcourt Brace, 1973.

———. *Home Economics.* New York: North Point, 1987.

———. *The Unsettling of America: Culture & Agriculture.* San Francisco: Sierra Club Books, 1977.

Booth, Annie L., and Harvey L. Jacobs. "Ties That Bind: Native American Consciousness as a Foundation for Environmental Consciousness." *Environmental Ethics* 12.1 (1990): 27–43.

Bornstein, Miriam. "The Voice of the Chicana in Poetry." *Denver Quarterly* 16 (1981): 28–47.

Campbell, SueEllen. "The Land and Language of Desire: Where Deep Ecology and Post-Structuralism Meet." *Western American Literature* 24.3 (November 1989): 65–79.

Chavis, Benjamin. "Foreword." *Confronting Environmental Racism: Voices from the Grassroots.* Ed. Robert D. Bullard. Boston: South End Press, 1993. 3–5.

Chrystos. "Anthropology." *Dream On.* Vancouver: Press Gang, 1991. 78–79.

———. *Fire Power.* Vancouver: Press Gang, 1995.

———. *Not Vanishing.* Vancouver: Press Gang, 1988.

Coltelli, Laura, ed. *Winged Words: American Indian Writers Speak.* Lincoln: U of Nebraska P, 1990.

———, ed. *Joy Harjo: The Spiral of Memory: Interviews.* Ann Arbor: U of Michigan P, 1996.

Cronon, William. *Changes in the Land: Indians, Colonists, and the Ecology of New England.* New York: Hill and Wang, 1983.

DeShazer, Mary K. "'Sisters in Arms': The Warrior Construct in Writings by Contemporary U.S. Women of Color." *Writing the Woman Artist: Essays on Poetics, Politics, and Portraiture.* Ed. Suzanne W. Jones. Philadelphia: U of Pennsylvania P, 1991. 261–286.

Diamond, Irene, and Gloria Feman Orenstein, eds. *Reweaving the World: The Emergence of Ecofeminism.* San Francisco: Sierra Club Books, 1990.

Dirlik, Arif. "The Postcolonial Aura: Third World Criticism in The Age of Global Capitalism." *Critical Inquiry* 20 (Winter 1994): 328–356.

DiZerega, Gus. "Individuality, Human and Natural Communities, and the Foundations of Ethics." *Environmental Ethics* 17 (Spring 1995): 23–37.

Driskell, Leon. "Wendell Berry." *Dictionary of Literary Biography* vol. 5. 62–65.

Eagleton, Terry, et al. *Nationalism, Colonialism, and Literature.* Minneapolis: U of Minnesota P, 1990.

Emberley, Julia V. *Thresholds of Difference: Feminist Critique, Native Women's Writings, Postcolonial Theory.* Toronto: U of Toronto P, 1993.

Evernden, Neil. "Beyond Ecology: Self, Place, and the Pathetic Fallacy." Glotfelty and Fromm 92–104.

Fast, Robin Riley. “Borderland Voices in Contemporary Native American Poetry.” *Contemporary Literature* 36.3 (1995): 508–536.

Gaard, Greta. “Ecofeminism and Wilderness.” *Environmental Ethics* 19 (Spring 1997): 5–24.

Gilreath, James. “Introduction.” *The North American Indian Portfolios from the Library of Congress.* New York: Abbeville, 1993. 6–26.

Glotfelty, Cheryll. “Introduction.” Glotfelty and Fromm xv–xxxvii.

——— and Harold Fromm, eds. *The Ecocriticism Reader: Landmarks in Literary Ecology.* Athens: U of Georgia P, 1996.

Griffin, Susan. *Woman and Nature: The Roaring Inside Her.* NewYork: Harper & Row, 1978.

Hall, Stuart. “Cultural Identity and Diaspora.” Williams and Chrisman 392–403.

Harjo, Joy. *In Mad Love and War.* Middletown: Wesleyan UP, 1990.

———. “Ordinary Spirit.” *I Tell You Now: Autobiographical Essays by Native American Writers.* Eds. Brian Swann and Arnold Krupat. Lincoln: U of Nebraska P, 1987. 269.

———. *The Woman Who Fell from the Sky: Poems.* New York: Norton, 1994.

Henley, Joan. “Exploring the Ways to Rainy Mountain.” *Approaches to Teaching Momaday's* The Way to Rainy Mountain. Ed. Kenneth M. Roemer. New York: MLA, 1988. 47–53.

Hernández-Ávila, Inés. “Relocations Upon Relocations: Home, Language, and Native American Women's Writings.” *American Indian Quarterly* 19.4 (Fall 1995): 491–507.

Hogan, Linda. *The Book of Medicines.* Minneapolis: Coffee House, 1993.

———. *Dwellings: A Spiritual History of the Living World.* New York: Touchstone, 1995.

———. *Mean Spirit.* New York: Ivy Books, 1990.

———. *Power.* New York: W. W. Norton, 1998.

———. *Solar Storms.* New York: Scribner, 1995.

Ignatow, David. Review of Wendell Berry's *Clearing. Partisan Review* 44.2 (1977): 317.

Jewett, Sarah Orne. *The Country of the Pointed Firs and Other Stories.* 1896. New York: Anchor Books, 1989.

Kaplan, Amy. "'Left Alone with America': The Absence of Empire in the Study of American Culture." Eds. Kaplan and Donald Pease. *Cultures of United States Imperialism.* Durham: Duke UP, 1994. 3–21.

Kaplan, Caren. "Deterritorializations: The Rewriting of Home and Exile in Western Feminist Discourse." *The Nature and Context of Minority Discourse.* Eds. Abdul R. Janmohamed and David Loyd. New York: Oxford UP, 1990. 357–68.

King, Ynestra. "Healing the Wounds: Feminism, Ecology, and the Nature/Culture Dualism." Diamond and Orenstein 120.

Kolodny. Annette. *Lay of the Land: Metaphor as Experience and History in American Life and Letters.* Chapel Hill: U of North Carolina P, 1975.

Kowalewski, Michael. "Writing in Place: The New American Regionalism." *American Literary History* 6.1 (1994): 171–83.

Larson, Sidner. "Native American Aesthetics: An Attitude of Relationship." *MELUS: Multi-Ethnic Literature of the United States* 17.3 (1991-92): 53–67.

Littlefield, Daniel F. "American Studies, American Indian Scholars, and the American Literary Canon." *American Studies* 33.2 (Fall 1992): 95–111.

Love, Glen A. "*Et in Arcadia Ego:* Pastoral Theory Meets Ecocriticism." *Western American Literature* 27 (1992): 195–207.

———. "Revaluing Nature: Toward an Ecological Criticism." Glotfelty and Fromm 225–240.

Manley, Kathleen, and Paul W. Rea. "An Interview with Simon Ortiz." *Journal of the Southwest.* 31.2 (Autumn 1989): 362–377.

McClintock, Anne. "The Angel of Progress: Pitfalls of the Term Post-Colonialism." Williams and Chrisman 291–304.

McKenney, Thomas, and James Hall. *History of the Indian Tribes of North America.* Philadelphia: Daniel Rice and James G. Clark, 1837.

Minh-ha, Trinh T. *Woman, Native, Other: Writing Postcoloniality and Feminism.* Bloomington: Indiana UP, 1989.

Mishra, Vijay and Bob Hodge. "What Is Post(-)colonialism?" Williams and Chrisman 276–290.

Mohanty, Chandra Talpade. "Under Western Eyes: Feminist Scholarship and Colonial Discourse." Williams and Chrisman 196–220.

Momaday, N. Scott. *The Way to Rainy Mountain.* Albuquerque: U of New Mexico P, 1969.

Murphy, Patrick. *Literature, Nature, Other: Ecofeminist Critiques.* Albany: State U of New York P, 1995.

Nash, Roderick. *Wilderness and the American Mind.* Third ed. New Haven: Yale UP, 1967, 82.

Ortiz, Simon. *Fight Back: For the Sake of the People, For the Sake of the Land.* Ortiz, *Woven Stone* 285–365.

———. *Going for the Rain.* Ortiz, *Woven Stone* 35–148.

———. *A Good Journey.* Ortiz, *Woven Stone* 149–284.

———. "Introduction." Ortiz, *Woven Stone* 3–33.

———. "Towards a National Indian Literature: Cultural Authenticity in Nationalism." *MELUS: Multi-Ethnic Literature of the United States* 8.2 (Summer 1981): 7–12.

———. *Woven Stone.* Tucson: U of Arizona P, 1992.

Owens, Louis. "Afterword." *Bearheart: The Heirship Chronicles.* By Gerald Vizenor. Minneapolis: U of Minnesota P, 1978. 247–254.

Papovich, J. Frank. "Journey into the Wilderness: American Literature and *The Way to Rainy Mountain.*" *Approaches to Teaching* The Way to Rainy Mountain. Ed. Kenneth M. Roemer. New York: MLA, 1988. 116–123.

Perrin, Noel. "Forever Virgin: The American View of America." *On Nature: Nature, Landscape, and Natural History.* Ed. Daniel Halpern. San Francisco: North Point Press, 1987. 13–22.

Platt, Kamala. "Ecocritical Chicana Literature: Ana Castillo's 'Virtual Realism.'" *ISLE: Interdisciplinary Studies in Literature and Environment* 3.1 (Summer 1996): 67–96.

Ponzio, Augusto. "The Relation of Otherness in Bakhtin, Blanchot, Levinas." *RSSI: Recherches Semiotiques Semiotic Inquiry* 7 (1987): 1–17.

Roemer, Kenneth. "The Heuristic Powers of Indian Literature: What Native Authorship Does to Mainstream Texts." *SAIL: Studies in American Indian Literatures* 3.2 (Summer 1991): 8–21.

Rose, Wendy. *Bone Dance: New and Selected Poems, 1965–1993*. Tucson: U of Arizona P, 1994.

Scarry, John. "Representing Real Worlds: The Evolving Poetry of Joy Harjo." *World Literature Today* 66 (1992): 286–291.

Schweninger, Lee. "Writing Nature: Silko and Native Americans as Nature Writers." *MELUS: Multi-Ethnic Literature of the United States* (1993): 47–59.

Shiva, Vandana. "Development as a New Project of Western Patriarchy." Diamond and Orenstein 121.

Silko, Leslie Marmon. "Landscape, History, and the Pueblo Imagination." *The Ecocriticism Reader: Landmarks in Literary Ecology*. Eds. Cheryl Glotfelty and Harold Fromm. Athens: U of Georgia P, 1996. 264–275.

Snyder, Gary. *A Place in Space*. Washington D.C.: Counterpoint, 1995.

Velie, Alan. "Gerald Vizenor's Indian Gothic." *MELUS: Multi-Ethnic Literature of the United States* 17 (Spring 1991/1992): 75–85.

Vizenor, Gerald. *Bearheart: The Heirship Chronicles*. Minneapolis: U of Minnesota P, 1978.

———. *The People Named the Chippewa*. Minneapolis: U of Minnesota P, 1984.

White, Hayden. "The Historical Text as a Literary Artifact." 1974. *Critical Theory Since 1965.* Eds. Hazard Adams and Leroy Searle. Tallahassee: U of Florida P, 1986. 395–407.

Wiget, Andrew. *Native American Literature.* Boston: Twayne, 1985.

Williams, Patrick, and Laura Chrisman, eds. *Colonial Discourse and Post Colonial Theory: A Reader.* New York: Columbia UP, 1994.

Yarbro-Bejarano, Yvonne. "Gloria Anzaldúa's *Borderlands/La Frontera:* Cultural Studies, 'Difference,' and the Non-Unitary Subject." *Cultural Critique* 28 (Fall 1994): 5–28.

Young, Vernon. Review of *Clearing* by Wendell Berry. *Contemporary Literary Criticism* Vol. 27. Detroit: Gale Research Company, 1984. 35–36.

Index

E

F

G

H

I

J

K

L

M

N

O

P

R

Y

AMERICAN INDIAN STUDIES

Elizabeth Hoffman Nelson and Malcolm A. Nelson
General Editors

The American Indian Studies series represents a growing group of important books on the literatures and cultures of America's indigenous peoples. The series is inclusive and open to a wide variety of approaches. We welcome scholarly literary studies and interdisciplinary studies of languages and cultures by American Indians, First Nations writers, and non-American Indians.

Original primary texts by American Indian and First Nations authors, thinkers, and religious and political leaders are especially encouraged.

For the submission of manuscripts, contact:

Heidi Burns, Senior Editor
Peter Lang Publishing, Inc.
P.O. Box 1246
Bel Air, MD 21014-1246
e-mail: hburnsplp@aol.com

To order other books in this series, please contact our Customer Service Department:

(800) 770-LANG (within the U.S.)
(212) 647-7706 (outside the U.S.)
(212) 647-7707 FAX

Or browse online by series:

www.peterlangusa.com